Going Underground

The Science and History of Falling through the Earth

Other Related Titles from World Scientific

The Pull of History: Human Understanding of Magnetism and Gravity through the Ages
by Yoshitaka Yamamoto
ISBN: 978-981-3223-76-9

A Dialogue Concerning the Two Chief Models of Planet Formation
by Michael Mark Woolfson
ISBN: 978-1-78634-272-0
ISBN: 978-1-78634-273-7 (pbk)

Newton and Modern Physics
by Peter Rowlands
ISBN: 978-1-78634-329-1
ISBN: 978-1-78634-330-7 (pbk)

Cooking Cosmos: Unraveling the Mysteries of the Universe
by Asis Kumar Chaudhuri
ISBN: 978-981-3145-76-4
ISBN: 978-981-3145-77-1 (pbk)

Solving Everyday Problems with the Scientific Method: Thinking Like a Scientist
Second Edition
by Don K Mak, Angela T Mak and Anthony B Mak
ISBN: 978-981-3145-29-0
ISBN: 978-981-3145-30-6 (pbk)

Going Underground
The Science and History of Falling through the Earth

Martin Beech
Campion College, The University of Regina, Canada

World Scientific

NEW JERSEY · LONDON · SINGAPORE · BEIJING · SHANGHAI · HONG KONG · TAIPEI · CHENNAI · TOKYO

Published by

World Scientific Publishing Co. Pte. Ltd.

5 Toh Tuck Link, Singapore 596224

USA office: 27 Warren Street, Suite 401-402, Hackensack, NJ 07601

UK office: 57 Shelton Street, Covent Garden, London WC2H 9HE

Library of Congress Cataloging-in-Publication Data

Names: Beech, Martin, 1959– author.

Title: Going underground : the science and history of falling through the Earth / Martin Beech (The University of Regina, Canada).

Description: Singapore ; Hackensack, NJ : World Scientific Publishing Co. Pte. Ltd., [2019] | Includes bibliographical references and index.

Identifiers: LCCN 2018049797| ISBN 9789813279032 (hardcover ; alk. paper) | ISBN 9813279036 (hardcover ; alk. paper) | ISBN 9789811201288 (pbk ; alk. paper) | ISBN 9811201285 (pbk ; alk. paper)

Subjects: LCSH: Science--History. | Mathematics--History. | Philosophy and science--History. | Thought experiments.

Classification: LCC Q174.8 .B42 2019 | DDC 509--dc23

LC record available at https://lccn.loc.gov/2018049797

British Library Cataloguing-in-Publication Data

A catalogue record for this book is available from the British Library.

For any available supplementary material, please visit
https://www.worldscientific.com/worldscibooks/10.1142/11236#t=suppl

Desk Editor: Ng Kah Fee

Typeset by Stallion Press
Email: enquiries@stallionpress.com

Printed in Singapore

Contents

Introduction

The well in which Truth is said to reside is really a bottomless pit.

Electromagnetic Theory (Vol. 1), Oliver Heaviside (1893).

History, the dour Scottish philosopher Thomas Carlyle once remarked, is the distillation of rumor. Nowhere is this better demonstrated than in the story of Galileo and the Leaning Tower of Pisa. Open any recently published book on physics and you will find some version of the events described by Galileo's first biographer and former student Vincenzo Viviani. Writing in 1654 Viviani describes a grand and flamboyant experiment, supposedly conducted in 1591, in which the then youthful mathematics professor at the University of Pisa, Galileo Galilei, climbed the more than 290 steps to the top of the city's famous Leaning Tower and from there threw-off from its upper ledge two spheres. One sphere was of a much greater weight than the other, and yet, and in spite of the received wisdom of the renowned ancient Greek philosopher Aristotle, the two weights, to the utmost surprise of the assembled students below, hit the ground at exactly the same time — as near as the eye could judge. It was a startling result and one that ran entirely counter to Aristotelian teaching. Indeed, Aristotle had taught that heavier objects should fall more rapidly than lighter ones in direct proportion to their mass. Imagine, then, that Galileo (or more likely some hapless student) had carried a 10 kg and a 1 kg mass to the top of the Leaning Tower. Since the two weights would be falling through the same distance, the height of the tower (a distance of about 55 meters), Aristotle would have argued that in the time that it took the 10 kg weight to strike the ground, the 1 kg weight would have fallen just $1/10^{th}$ of the way (about 5.5 meters) down the tower's side. In triumph Galileo had demonstrated, since the two weights struck the ground at the same time, that Aristotle's ancient description of motion must be wrong. In modern terminology, and this is why it is found in present-day physics books, we would say that Galileo had demonstrated that all objects, irrespective of their mass, fall with the same acceleration. The Pisa experiment described by Viviani has all the hallmarks that would be expected of Galileo — it was showy, public, in your face, simple, elegant, and in complete defiance of accepted wisdom. The only problem with Viviani's account is that no such experiment by Galileo ever took place — it is a complete and utter fabrication.

We can be sure, through multiple lines of reasoning, that Galileo never performed the experiment accredited to him by Viviani. Principally, we know enough about Galileo's extended ego that if he had performed such an experiment then he would most definitely have made sure that everybody knew about the result. And yet, when he produced his great work on dynamics and the material properties of matter, *A Discourse on Two New Sciences*, published in 1638, Galileo remains silent with respect to any experimental verification of how objects of different mass actually fall. Certainly he presents the argument that objects fall with the same acceleration irrespective of their mass, but he presents the result as a thought experiment. Indeed, Galileo sets up a simple thought experiment to reveal a contradiction in the Aristotelian expectation. Imagine, Galileo argues, that two unequal mass weights are held together by a piece of string, then we could imagine two possible outcomes if they are let fall through some fixed distance. In the first case we could imagine that the lighter weight, traveling more slowly than its heavier companion, would retard the motion and accordingly slow the whole system down. In this case the two objects will take longer to hit the ground than the heavier weight let fall on its own. In contrast, we could also argue that by tying the two weights together the total mass of the falling object is increased, and therefore it should hit the ground sooner than the more massive weight on its own. By straightforward logic a contradiction has been revealed — the combined, string-tied weight falls both more slowly and more rapidly than the individual components. This contradiction,

Galileo correctly reasoned argues against weight (or more correctly mass) as having anything to do with the speed with which objects fall. This result, a fundamental result in fact of dynamics, was found through reason alone, it was a thought experiment that Galileo performed, and it was not predicated upon the performance of any actual experiments. This kind of thought experiment today is an essential part of the theoretician's tool kit. It is a valid and accepted means of exploring the possible consequences or outcomes of hypothetical question that may, or may not, be open to direct experimentation. Indeed, it is the importance of Galileo's reasoned result that a body falls at a rate independent of its mass that explains why it is described in all present-day physics texts, some 400 years onwards in time. And the Leaning Tower of Pisa bit? Well, that just adds color and interest to the story; it is the sort of experiment that we would like to think Galileo performed.

Through the power of thought alone Galileo reasoned his way to a contradiction about how objects should fall, and this contradiction argued against the accepted wisdom, handed down by Aristotle, that heavier objects fall faster than lighter ones. But wait a minute, we might now ask, for the simplicity of the procedure described by Viviani, why didn't Galileo, or one of his students, actually perform the experiment as described. The answer here is probably that Galileo knew full well that any such experiment would probably fail. The failure, however, would not be because the thought experiment was wrong but because the experimental conditions could not be controlled and because the measurements themselves would be very difficult to make — at least in 1591. To make the experiment work one would have to ensure that the various weights were released at exactly the same time, since any delay in releasing one weight would affect the impact comparison on the ground — recall that the test was to see if they hit the ground at the same time as witnessed by human observers. While mechanical clocks did exist in Galileo's time, none would have then had a seconds-hand, and certainly none could measure time intervals corresponding to just a fraction of second. Not only this, the effects of wind speed would need to be considered, since the force of the wind and its associated drag effects will affect the dynamics of each mass differently. In the modern era all these conditions can be controlled in the laboratory, but this was not an option open to Galileo. Indeed, the very first public demonstration of Galileo's thought experiment was not to be realized until 1971, when astronaut David Scott, during the Apollo 15 lunar landing mission, simultaneously dropped a feather and a hammer at the Moon's surface. In the vacuum of space and in the lower gravity environment of the Moon, the two objects, once released from Scott's hands, gracefully sank towards the Moon's surface in wonderful synchronicity. The experiment was beautifully simple, and simply beautiful, and it visually vindicates the intellectual power of Galileo's reasoning [1]. The 19th century Austrian physicist and philosopher Ernst Mach would have greatly appreciated the Apollo 15 result, since the realization of the Moon experiment cut to the core of his ideas concerning thought experiments. Mach specifically introduced the German term *gedankenexperiment* to denote the process by which a researcher must first imagine exactly how an experiment will work, and what results it will reveal, prior to actually performing the real physical experiment.

While Galileo's thought experiment stopped when the various weights hit the ground, the thought experiment that is the main focus of this text asks a somewhat different question. Our thought experiment essentially begins where Galileo's left-off. What, we ask, would happen if those same weights, dropped from the top of the tower at Pisa, just kept on going? Rather than being compelled to stop at the Earth's surface, what would the path and speed of those same weights be if, like some mysterious ghost matter, they could move through Earth's interior unencumbered by the presence of physical matter, but still subject to the gravitational influence of that matter. In essence the thought experiment to be considered is that of Earth-tunneling. This thought experiment is no flight of fantasy, which is not to say that its consideration hasn't resulted in the fantastical. It is, rather, a device, or question of the imagination that has been used on numerous occasions throughout recorded history to investigate the nature of the Earth's interior and to probe the fundamental laws of dynamics.

The Earth-tunneling thought experiment, as we shall see in subsequent chapters, both pre- and post-dates Galileo. It was a problem considered by Aristotle circa 250 B.C.; who used its solution to reason that the Earth must be spherical in shape (see Chapter 10). It was also part of a fundamental question and partial experiment considered by Robert Hooke and Isaac Newton in 1680 (see Chapter 13). And, indeed, it was the subsequent vitriol and venom that developed between Hooke and Newton, in no small part relating to the Earth-tunneling thought experiment, that was responsible for the eventual development of Newton's *Principia Mathematica* in 1687 — the most influential science book ever written (Chapter 14). Other philosophers have found the mathematical pathology of unbounded infinity within the Earth-tunneling thought experiment (see Chapter 16); while yet others have found the intriguing hare-and-hound paradox of Zeno within its compass (see Chapter 11). And, remarkably, those more mechanically inclined practitioners have used the Earth-tunneling thought experiment to attempt Mach's *gedankenexperiment* ideal — a result that in at least one case (see Chapter 16) resulted in the resignation from a prestigious Presidential post and an untimely and embittered death.

In addition to encountering Galileo's fictitious Leaning Tower of Pisa experiment, many present-day students, at some stage or another, are asked to confront its idealized mechanical realization in the classic Earth-tunneling problem. In this latter situation, rather than considering the passage of ghost matter through the Earth's interior, the problem is built upon the impossible premise of drilling a hole all the way through the Earth and then dropping a mass through the hole. The question, indeed, the very question that occupies our attention in Chapters 3, 4 and 5, is to describe the motion of the object let fall through the Earth-crossing tunnel's gaping maw. This specific problem has its own fascinating history, as will be revealed, and its connection with the gravity-train concept is a topic that has driven both Hollywood-style speculation and exploration through literary fiction (see Chapters 9, 17 and 18).

Chapter 1

A Long Distraction

What we call the beginning is often the end
And to make an end is to make a beginning.
The end is where we start from

T. S. Eliot

I had in mind a grand survey. The idea was to review the entire series of articles, from issue 1 to issue 3212 of *The English Mechanics and the World of Science*, and to thereby produce a great synthesis of the early days of science popularization [2]. For indeed, the time interval over which the *Mechanics* was published, 1865 to 1926, encompasses a period of great industrial development and social change. Enveloping a time span incorporating the first Great War, the invention of the telegraph, the radio, the motorcar, the development of spectroscopy, the introduction of special and general relativity, along with the growth of railway networks, the rise of assembly-line mass-manufacturing and the growth of organized scientific enquiry, the *Mechanics* oversaw the appearance of numerous life-changing and indeed, life-shattering revolutions. Somewhere along the path to producing my grand survey, however, I became distracted, and this book is a result of that distraction.

The English Mechanics and similar such magazines were the 19th century equivalent of today's internet. They offered news items, society-meeting updates, review articles and a notice-board-format (equivalent to the modern-day *chat room*) through which readers could raise questions, ask for help with problems and criticize and/or praise earlier articles and letters. The return time on questions was certainly slow when compared to the modern-day expectations, but typically within a few weeks some reader from somewhere would offer their sage advice on a problem that had been proposed. The download and uplink times may well have been glacial (again by modern standards) but the bandwidth was phenomenal. The topics discussed in the *Mechanics* readers *Queries and Answers* section were incredibly diverse, ranging from straightforward questions concerning steam boiler design, or car and bicycle maintenance, to specific mathematical features of Einstein's theory of general relativity. No topic was seemingly inappropriate to the magazine editorship, with historical articles on ancient Greek engineering sitting side by side with articles on the latest astronomical discoveries and articles about futuristic flying machines, walking automobiles, rockets and ships, and windmills driven by the Magnus Effect. There were articles on weather prediction, the structure of the Earth, microscopy, evolution theory, animal behavior, insects and photography. Ideas on spiritualism were discussed, along with new ideas for advanced submarine design, the formation of the universe, the transmission of radio-pictures around the globe, the possible detection of life on Mars, the future problems of feeding Earth's human population, and even prime number theory and Fermat's last problem. Indeed, after just a few minutes of reading through back issues of the *Mechanics* one cannot but find that there is very little that is really new and innovative taking place in the internet chat rooms of today — just the format has changed.

For all of the incredible diversity of topics presented within the *Mechanics*, in eye-popping, closely-spaced small type, the article that caught my eye, resulting in the writing of this book, was question number 128 on page 113 in the March 7th, 1924 issue (Figure 1.1). I will save the discussion of this question until Chapter 3, but it was a short, 5-line question, written by a reader identifying himself as Cymro, that set my mind running. The use of a pseudonym was not uncommon in the *Mechanics*, and similar such journals of the time and one can find letters and/or responses to letters by such regulars as *FRAS*, *Eos*, Sigma, Spen Valley, Jack of all Trades, The Harmonious Blacksmith, RAP, The Charge Hand, and NEB. In some cases the person and persona behind a pseudonym is completely unknown, but others became so well-known in their lifetime that their *nom de plumes* lost all anonymity. FRAS was Captain William

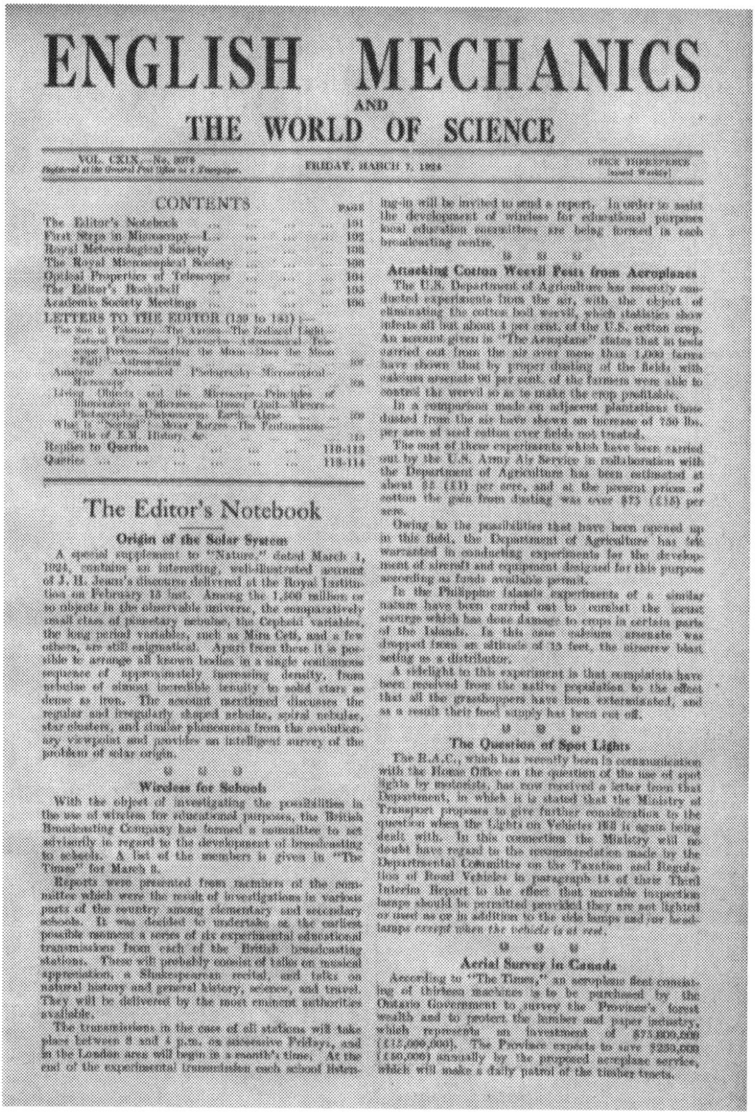

ENGLISH MECHANICS
AND
THE WORLD OF SCIENCE

FRIDAY, MARCH 7, 1924

CONTENTS

The Editor's Notebook

Origin of the Solar System

Wireless for Schools

Attacking Cotton Weevil Pests from Aeroplanes

The Question of Spot Lights

Aerial Survey in Canada

Figure 1.1. Front page of the *English Mechanics and The World of Science* for March 7th, 1924. The Editor's Notebook reveals that this particular issue contains a special supplement concerning James Jeans' ideas on the origin of the solar system, along with an article on the use of wireless radios in education, an article on the pesticide control of cotton weevils, an article on a new aerial survey of Canada, and a report on the use of headlights on automobiles.

Noble, a keen observer of meteors and a well-known Fellow of the Royal Astronomical Society; *Eos* was Major-General Lowther who provided prodigious amounts of information on how to make Indian chutney, the construction of canoes, jungle explorations and, indeed, the design of hernia trusses. RAP was Richard Proctor, a well-respected writer on all science topics and an early pioneer of science popularization; Jack of all Trades, turned out to be Mr. John Hollingsworth, an engineer and keen amateur gardener, who worked in the Royal Naval Dockyard at Sheerness in Kent, UK. The passage of time, now encompassing one hundred years, will presumably continue to preserve the anonymity sought by other writers and proposers of questions — Cymro is sadly amongst the list of unknown correspondents to *The English Mechanics* magazine. All that one might conclude about Cymro is that he was presumably from Wales, since *cymro* means Welshman or Welsh person in the Celtic language.

After the First Great War, readership of the *Mechanics* went into decline, indeed, many of its earlier readers were then dead from the four years of carnage that started in 1914. The magazine struggled on for a final decade but times were changing rapidly, and the dense type of the *Mechanics* along with its low-resolution reproductions of black and white photographs could not compete with the colorful glossies and the burgeoning entertainment industries which pumped out a continual salvo of new music and movie stars — the day of the well-informed general reader was coming to a close, and the days of the distracted, money-to-spend, thrill-seeking, consumer were beginning. For all the excitement and change, however, the *Mechanics* will have to wait a while longer before I complete my grand synthesis — one day.

In addition to Cymro's question of 1924, my attention was also drawn to a whole series of letters and articles that appeared over a time span of at least five years beginning circa 1920, concerning the structure of the Earth. What specifically surprised me was how little was actually known, at that time, about the Earth's interior — various august doctors and learned professors expounded upon the merits of various theories; theories that saw the Earth as being a solid cooled-off mass, through to that of a hyper-hot magma-filled interior surrounded by a rigid crust no more than a few miles thick. Seemingly every reader had a viewpoint, and they wanted their opinions heard (or at least read). Some rebuttals were firm and fair, others saw the authors bristle with umbrage. The Editor must have been thrilled at the content, and indeed to let the topic run for so many years there must have been plenty to write about. Arguments based upon pendulum experiments, the thermodynamics of heat flow, mining activities, historic and contemporary volcanic eruptions and earthquakes were all presented — even the lost world of *Atlantis* saw a mention. But what was missing from the debate was raw data, apart from mining and tunneling experiences, no one could actually say that they had been deep inside the Earth, or even conducted an experiment deep inside the Earth. This same argument, remarkably, applies to this very day (although see Chapter 19), and while we may now be able to map out the Earth's interior with seismic arrays, more humans have walked on the Moon (the so-called *Dusty Dozen* [1]) than have descended further than fifteen kilometers below the Earth's surface or sea-level (the number in the latter category actually being zero).

Humans have long exploited the resources of the Earth's surface, the oceans and the air, but the ground, the deep ground, the hallowed ground beneath our feet has been more the preserve of the dead than the explorer. The ground below us is solid and dependable — it is the undeniable boundary, the topological floor, from which humanity looks outwards into the vastness of the universe: only the hard-worked miner and the decaying bodies of our ancestors find work and peace below ground. The deep undersurface regions have, throughout human history, been associated with the covens of Hell and the repellent malevolence of the underworld, where only deviants and demons can hope to thrive and survive. To peer into the mouth of a cave, even to this very day, fills the mind with adventure and fear — fear of the darkness, fear of that which cannot be seen, and fear of becoming trapped within the winding and stone-lined corridors of an alien world. Caves and mines are oppressive, the weight of the world presses down upon one's soul, and they are where danger lurks, waiting and unseen, ready to trap and destroy any unprepared interloper. From the lost beginnings of the timeline of history, the explorer has eulogized, bragged and willfully-lied about the lands and mysteries that they have seen and experienced over the home horizon, but few have ever dared to venture underground. Only in the mind's eye has humanity ventured thither.

Chapter 2

Wishing Well

What mystery pervades a well!
That water lives so far –
Whose limit none has ever seen,
But just his lid of glass –
Like looking every time you please
In an abyss's face!

Emily Dickinson

It is a compulsion. Indeed, as one approaches the cylindrical surround of any water well the idea begins to build and build until it literally bursts into action: a small stone or coin must be dropped into the opening. With trepidation the head leans over the well opening and the walls diffuse into an inky blackness — somewhere, down there, in that darkness, is a body of water. The hand, poised and ready over the opening, releases its smooth pebble cargo. Through the silence the mind begins to count: one Mississippi, two Mississippi, three — ah, there it is, the sound of a distant splash. With an arrogance of certainty that only science can bring about, the depth of the well has been probed. In the first second the stone will fall about 5 meters; in the second second the stone will travel a further 15 meters; in the third second the downward plunge of the pebble will carry it a further 25 meters. In my mind's eye the picture, the abstract picture, is clear, and the motion is entirely Newtonian under constant gravity. If I hear the sound of the pebble splash 2 seconds after it is released from my hand, then the well is 20 meters deep. End of story. The certainty of this simple result, however, builds upon an incredible set of

ideas and abstractions, none of which being necessarily obvious —
unless, of course, you are a taciturn genius like Isaac Newton.

Isaac Newton (1643–1727), in spite of the writings of countless scholars and explainers over these past several centuries, remains a mysterious figure. More than human, he seems mythical, otherworldly and
hardly common at all. Certainly he did not lack for an ego; he could carry
a grudge; and being fully aware of his genius he was both paranoid and
self-grandiloquent. Newton, however, had a gift, a very rare gift, that has
been granted to but a minuscule few of all the human beings that have
ever lived. He had a mind that could grab a problem, wrestle it to the
ground, shake it apart, and re-assemble it with deep and meaningful
insight. His mind was so subtle in action that it could notice the obvious,
and then, having seen the same things that every other human had seen,
find the deep workings of nature at play. With all the self-modesty he
could muster, in later life, Newton suggested that "to myself I seem to
have been only like a boy playing on the seashore, and diverting myself
in now and then finding a smoother pebble or a prettier shell than ordinary, whilst the great ocean of truth lay all undiscovered before me." If
ever there was a sound bite worthy of all science, contrived and yet full
of meaning and depth, Newton's seaside story must surely be it.

Newton's greatest work, *Philosophiæ Naturalis Principia Mathematica*
(that is, *The Mathematical Principles of Natural Philosophy* — usually
referred to as the *Principia*), was published in 1687 — it was a work predicated on genius, channeled anger, distain and mysticism. The genius was
self-evident, and the results of Newton's intellectual struggles will be
discussed throughout the remainder of this book. The antagonism with
Hooke arose from an exchange of letters, starting in November 1679,
concerning the motion of a "bullet" falling through the Earth (as will be
further explored in Chapter 13). Newton, it transpired, made a mistake in
his analysis, and Hooke seeing an opportunity to score points over his
rival in Cambridge publicly pointed out the mistake. To Newton this was
almost unbearable, but importantly he channeled his anger into finding
the correct solution to a problem — and then he told no one about his
triumph. This secret solution, however, eventually resulted in the writing
of the *Principia* and it ushered in the concept of universal gravitational
attraction. The distain behind the *Principia* was for the reader, although

one might as well call a spade a spade and accept that the distain was mostly for Robert Hooke. Newton wrote to Edmund Halley (who financed, edited, and saw the *Principia* through the press), "to avoid being baited by little smatterers in mathematics I shall not stint in complexity."

The mysticism of the *Principia* concerned the very introduction of universal gravitational attraction — this was, in fact, the central doctrine of Book III, and probably Newton's best known contribution to science in the minds of the lay reader. Why is the idea of gravity mystical? Well, just look around you — do you see anything with your eyes that suggests objects are exchanging forces? Yes, things move or fall, or roll about, but there is no tangible medium or presence, that the human body can sense, to *drive* the motion that is observed. This latter argument, of course, is a false deduction, since just because we as human beings cannot see something does not mean that that something doesn't exist. It is this very issue that Newton addresses in the title of the *Principia* in that it is the mathematical principles of nature that he is concerned with. With respect to the question of what gravity is, Newton famously replied *hypothesis non fingo* (I fain no hypothesis). In essence Newton tells us in the *Principia* how gravity operates, through a set of specific mathematical relationships, but he does not tell us what gravity is — indeed, we still do not know what gravity is and why it exists as an operational agent within the observable universe. An increasing number of physicists and mathematicians in the modern era have taken Newton's stance even further, arguing that the universe itself is a mathematical structure, and that all we in fact know about anything is that which is expressible through mathematics. This, of course, is a troubling philosophy for many reasons and an approach, in the author's opinion, that is entirely misguided — but so be it.

In the *Principia*, Newton gives us equations that describe the motion of objects under varying forces. The force of gravity is introduced in Book III of the *Principia* as "there is a power of gravity pertaining to all bodies, proportional to the several quantities of matter which they contain." To this he later adds, "the force of gravity towards the several equal particles of any body is inversely as the square of the distance of places from the particles." These two statements combine to give us the now familiar formula for the gravitational force: $F_{gravity} = GMm/R^2$, where the M and m terms account for the masses of the two objects which are situated a distance *R* apart between their centers, and *G* is the universal gravitational constant. This is an idealized formula, with all extraneous details stripped away. We do not need to know what the objects are made of, stone, lead, iron or wax, or their color, or their physical shape. All we need is the mass in kilograms (which of course, for a given size, depends upon what the object is made of), and the separation of the objects — and even in this latter case the formula assumes that we know where the center of each object is: Newton takes the objects to be uniform density spheres, or at least sphere-like, but nature knows of no such objects. Nature produces objects that have non-uniform distributions of matter within non-spherical bodies. For all this, however, the beauty of Newton's abstractions and the shucking of extraneous detail is that the formula works — it gives the right answer when tested against actual measurements. And this is one of Newton's key points in the *Principia*, to understand how nature works we need to be guided by experimentation, actual numbers derived with actual measuring devices, to the formulation of idealized equations, sanitized of all unnecessary detail. It is this latter step, the formulation of the idealized equations, that requires human imagination to literally take flight and embrace the thought experiment — that is to visualize and explore with the mind's eye the "what if world." What if we can drill a hole through the Earth? What if we drop a stone through that tunnel? What if the world is a spherical shell or what if a stone let go from a high tower can move unimpeded through Earth's interior rather than stopping abruptly at its surface?

Newton developed his three laws of motion around the physical experiments conducted by Galileo Galilei. These laws tell us how objects move in an idealized universe: a universe infinite in extent and a universe that is (at least potentially) infinitely old. They ignore extraneous events and they tell us what will happen when forces act, or do not act, upon an object. Newton's first law of motion explains that, an object in a state of rest or uniform motion remains in that state unless acted upon by some net external force. This law begins to state the obvious, although the universe doesn't work according to the ideas of what human beings find obvious. An object at rest will remain at rest, well that's obvious, but what Newton is saying is that it will remain at rest forever — and forever is a very long time. Likewise, if an object is moving with uniform motion, then what Newton is telling us is that it will do so forever and it will never stop and it will never veer away from a straight line trajectory. Nothing in a finite-aged and finite-sized universe containing matter (gas, dust, stars and planets) can actually achieve such ideals, but in a thought experiment or the mind's eye world, these conditions not only hold true, they are perfectly reasonable statements to make. Under the philosophy outlined by Greek philosopher Plato such laws are ideals. The ideal forms exist and are absolutely true *somewhere*, but they are only approximately true in the world that we find ourselves actually living in [3]. They are only approximately true in the everyday world of our human senses since the world we experience, warts and all, is messy: there is no such entity as a perfect sphere, there is no motion without some form of friction, and there is no such thing as a constant force, and so on. In the real world, objects set in motion slow down after some time, or hit a wall, or get wet, or change shape — indeed, they are continually affected by a myriad of small, random perturbations that push and distort the mathematical certainty of the idealized world into the contorted and chaotic world that we, as human beings, experience. The laws of motion are real and true statements in the mind's eye (or thought experiment) world, and they approximately map onto the physical world around us — this was (and remains) an incredible realization, and it is the mapping part of the process that enables mathematics to provide us with astonishingly good answers (in some cases) concerning the working of the universe. For all this, however, the universe is not a mathematical structure. Mathematics is simply the (non-unique) bridging function between the ideal, thought-experiment world, and the everyday complex world in which we live. For all this, numbers and mathematical proofs provide us with absolutes (unreal as they are). Indeed, while no real object can experience the idealized motion described by the words in Newton's first law, mathematical theorems can be constructed, and proved correct, within the framework of mathematics, with the correctness of the theorems literally holding forever. Mathematical proofs are truly idealized, everlasting constructs, and we may be certain, for example, that at no time between now and the very end of time (in the far, far distant future) that no counter argument will ever be found to negate the correctness of the Pythagorean Theorem in Euclidian space (Figure 2.1) [4].

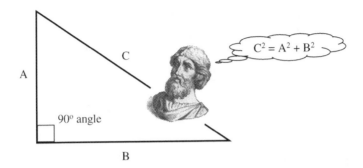

Figure 2.1.   A picture of eternal certainty and ever-lasting truth. At no time, even into the most remote, far future, will the Pythagorean Theorem be negated: C^2 will always equal $A^2 + B^2$ in a perfect right-angle triangle.

Mathematical theorems, if (correctly) proven true once, remain true forever. The real world has no such parallel, and we are truly fortunate, therefore, that some of the certainty from the idealized mathematical world occasionally seeps through into the real world, and thereby offers us a guide as to how the real world sometimes works. As has been said by many others before now, the real world is unreasonably well described by mathematics. All that I am really trying to say at this stage in our narrative is that thinking about drilling a hole through the Earth, and then dropping an object into that tunnel is no more bizarre a question in the thought-experiment universe than asking, the seemingly simpler question, what is the motion of a sphere as it rolls down a hill, or for that matter, asking how one object moves under the gravitational influence of another.

Newton's second and third laws are more utilitarian in practice, and they enable us to say something about the way in which forces are made visible (through the motion of objects) and how objects respond to the forces impinging upon them. The first law tells us that if an object moves along some curvy path then it must be experiencing some net force that is pulling it away from a straight-line path. Likewise, even if an object is moving in a straight line, but does so with a changing speed, then it too must be suffering some net force that is driving the acceleration. Newton's second law, in fact, tells us that if some net force F acts upon an object then the acceleration, or change in velocity, *a* that it will produce is inversely proportional to the mass *m* — that is, in symbolic form, $a = F/m$. So for a given net force *F*, a massive object will suffer a smaller acceleration than that experienced by a lower-mass object.

The third law is a statement about the conservation of motion, and it essentially says that the total amount of motion in the universe is constant at all times. While the total amount of motion is constant, however, it can be transferred between objects. The third law is expressed in terms of action, and says, for every action there is an equal and opposite reaction. What this means is that in every interaction between two objects there are a pair of forces in operation, with the size of the force on one object being equal but directed in the opposite direction to the force acting upon the second object. The importance of the third law is that it leads directly to the more fundamental result of the conservation of momentum [5].

It was Galileo in his *Dialogue Concerning Two New Sciences*, published in 1638, who first described mathematically and verified experimentally how an object moves under the condition of constant acceleration. Guided by a series of experiments in which various sized spheres were rolled down an inclined plane and adapting the theoretical reasoning of the Merton calculators (see Chapter 11), Galileo reasoned that the distance *s* traveled by an object undergoing a constant acceleration *a* for a time *t*, will vary as the time squared, with symbolically: $s = \frac{1}{2}at^2$. If the body was moving with some constant velocity *u* before the acceleration began, then the total distance traveled in time *t* will be $s = ut + \frac{1}{2}at^2$, and this, under the condition of no acceleration ($a = 0$) yields the familiar, everyday equation relating the distance traveled in time *t* at constant velocity as: $s = ut$. Likewise, Galileo reasoned that the velocity *v* of a body undergoing constant acceleration over time *t* will be: $v = u + at$. These formulae successfully describe the distance traveled by an object and its velocity at a specific time in the case of constant acceleration, and this situation, of course, applies at the Earth's surface. Indeed, Newton additionally tells us in his *Principia* that the acceleration of any falling body (for example an orange dislodged from a tree) located close to the Earth's surface is $a = g = GM/R^2$, where *G* is the universal gravitational constant, *M* is the mass and *R* is the radius of the Earth.

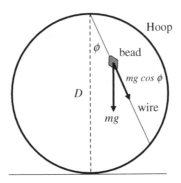

Figure 2.2.   A two-dimensional "hoop" analog of the Earth-crossing tunnel.

To work our way towards the solution of Cymro's Earth-tunnel question, let us first begin with a simpler constrained, constant acceleration problem. Imagine that we have a circular hoop and a thin piece of wire that is attached at one end to the very top of the hoop, and at the other to any location on the hoop's circumference (Figure 2.2). Now, imagine that a small bead of mass m is allowed to slide without friction along the wire. What is the travel time of the bead along the length of the wire from its top position to the bottom? The bead will travel along the wire under the influence of Earth's surface gravity g, and accordingly, $F_{gravity} = GMm/R^2 = gm$, where laboratory measurements indicate that $g = GM/R^2 = 9.81$ m/s^2. The force acting on the bead driving it along the wire will be $F = gm \cos \phi$, where ϕ is the angle that the wire makes to the vertical, and the resultant acceleration will be $a = g \cos \phi$. Now, the length of the wire L, which is a chord across the hoop, is given by $L = D \cos \phi$, where D is the diameter of the hoop.[1] Using Galileo's equation that links the distance traveled, the acceleration and the time t, where t will be our wire-crossing time, we have $L = \frac{1}{2} at^2 = \frac{1}{2} g (\cos \phi) t^2 = D \cos \phi$. We now obtain the somewhat interesting and perhaps unanticipated result that the bead travel time t_{hoop} along the wire is independent of the mass of the bead and that it is also independent of the angle ϕ of the wire to the vertical, and that $t_{hoop} = (2D/g)^{1/2} = 2 (R/g)^{1/2}$, where R is the hoop

---

[1] This is a consequence of Thales' Theorem, which dictates that the inscribed angle across any diameter of a circle must be a right-angle.

radius — as one might expect. This result indicates that the bigger the hoop the longer the travel time of the bead along the wire.

In the 2-dimensional hoop problem just considered, the gravitational driving force *mg* is taken as being constant and acting downward in a direction vertical to the ground. In the case of a solid object, and specifically in Cymro's problem, this will not be the case. And accordingly we need to ask where does the gravity of a solid object reside? According to Newton the gravitational force is always directed towards the very center of the object (like Newton, we shall assume that the objects being considered are always spheres — another mind's eye approximation). Indeed, Newton's fundamental formula is based upon the idea of point masses — that is, the gravitational attraction is determined as if all of the mass of an object is concentrated into an infinitely small point at its center. This, of course, is an odd notion since any physical object has matter distributed throughout its entire body. At a point inside of a sphere, Newton's shell theorem tells us that the gravitational force will depend upon the matter interior to the point's location only (see Figure 2.3). If we have a constant-density sphere of radius R and a point mass at some location $r < R$, the gravitational attraction on the point mass is dependent only upon the matter within the sphere of radius r. The matter located within the spherical shell between r and R has no net gravitational effect at all. This again is another remarkable result and it leads us to the corollary that there is no net gravitational force acting upon a point mass located within a spherical shell. In other words, the point mass will remain at rest, or in uniform motion, no matter where it is located within the spherical shell (unless it hits the shell's inside surface, of course, in which case it will bounce).

If we now go back to our 2-dimensional hoop problem and imagine spreading-out the hoop material to fashion it into a 3-dimensional thin shell, we may then ask, what does the bead attached to the thin wire do in the absence of any external gravitational field (say that of the Earth). The answer now is that the bead does nothing, nothing at all — it simply stays put. Under the gravitational influence of the shell alone, Newton's shell theory dictates that there is no net force acting upon the bead (or any object inside of the shell) and consequently it will remain in place for all time — now the travel time of the bead along the wire, if one likes to

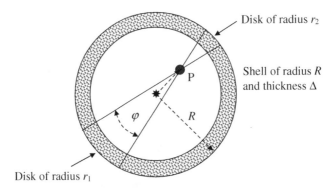

Figure 2.3. An outline proof of Newton's shell theory. Let the shell have a thickness Δ, a density ρ, and a radius R. Let also the particle P be located at a point within the shell at some distance r from the shell center. Constructing two cones, each with a vertex angle φ at P, two circular disks are traced out upon the surface of the shell. The mass enclosed within each circular disk will relate to the product of its volume times the density. The radius of disk 1 is simply $r_1 = \tan(\varphi/2)\,(R + r)$, and the radius of disk 2 is likewise, $r_2 = \tan(\varphi/2)\,(R - r)$. The gravitational force of disk 1 at P will be $F_1 = G\pi r_1^2 \Delta\rho/(R + r)^2$, and the gravitational force of disk 2 at P will be $F_2 = G\pi r_2^2 \Delta\rho/(R - r)^2$. Taking the ratio of these two forces we have, $F_1/F_2 = (r_1/r_2)^2[(R - r)/(R + r)]^2$ which upon substituting for r_1 and r_2 yields $F_1/F_2 = 1$. In other words, the gravitational force due to disk 1 and disk 2 acting on P are such that they are of the same magnitude, but because they act in opposite directions there is no net force acting upon P. We can now imaging constructing similar such cones about P to account for the gravitational influence of the entire shell, and in each case the gravitational attraction will cancel out. By this reasoning, we recover the result that there is no net gravitational force acting upon a particle P placed inside a hollow shell. It is important to note that this result is not the same as saying that there can be no gravitational field inside a hollow shell; rather it says that the shell itself does not contribute to the overall gravitational field (e.g., the Earth's gravitational field) within which the shell may sit.

think of it at all, is infinitely long. The only way in which we can make a particle move through a spherical shell is to give it some initial velocity V and then the travel time across the shell will be $t = D/V$, where D is the shell diameter. One way that this might be achieved (see the Appendix) is suggested in relation to a problem posed by Peter Tait and William Steele, in their wonderful text, *A Treatise on the Dynamics of a Particle* (first published in 1865). Here the idea is to let a small particle fall towards the Earth-shell from a great distance away so that upon entering the shell, through a small aperture, it will have the Earth-shell's so-called escape velocity [6]. Under these conditions the Earth-shell crossing time is $t_{shell} = 2R/V_{esc} = (2R/g)^{1/2}$, which is slightly smaller (by a factor of $2^{1/2}$) than the hoop crossing time derived earlier.

In terms of solving Cymro's problem we have found, so far, that the 2-dimensional hoop analog indicates a crossing time[2] $t_{hoop} = 2(R/g)^{1/2}$; the 3-dimensional shell analog with a small particle falling in from *infinity* is $t_{shell} = (2R/g)^{1/2} = t_{hoop}/2^{1/2}$. With these analogs, it turns out that we have found the correct relationship between the crossing time, the radius and the surface gravity (see the Appendix for full details), but in order to solve Cymro's problem exactly we need to dig a little deeper into the physics of the Earth, and indeed, introduce his actual problem.

---

[2] If we (bravely) insert numbers for the Earth's radius and surface gravitational acceleration into the travel time formula for the 2-dimensional hoop problem, then $t_{hoop} = 26.9$ minutes; the in-fall from *infinity* Earth-shell crossing time (see the Appendix) is just 19 minutes.

Chapter 3

Cymro's Problem

Ac yno yn y dyffryn tawel mi
Glywaf gân yn swn yr awel[1]

Ysbryd y Nos by E. H. Dafis

Readers of *The English Mechanics* for March 7th, 1924, found on page 113, in query number 128, of their weekly reading the following question:

Qu.128 — If two points on the surface of the Earth, taken to be a uniform sphere of radius *a*, were connected by a straight smooth tube, prove that a particle introduced into the tube at one extremity and then released would move in the tube with simple harmonic motion. In what time will the particle arrive at the other extremity of the tube; also what will be its velocity at the middle point of the tube? *Cymro*.

I knew the answer as soon as I saw the question on the reproduction of the magazine page in front of me — it's a classic. Then I paused. Yes, I knew the answer and as a first-year undergraduate student at the University of Sussex in 1975 I had struggled through its solution. Then it was a slog — now, many decades later, I think of it as a classic problem. What had changed? Well, of course, many things have changed since the mid-1970s: countless life-changing events have happened, outlooks have changed, and time has relentlessly marched by. Since 1975 I have encountered many variants of Cymro's problem — almost every recent introductory physics text and/or tome on ordinary differential equations carries the question in one form or another, and in many ways it has

---

[1] From the Welsh: *And there in the quiet valley I hear a song in the sound of the breeze.*

19

become a right-of-passage[2] kind of question. None of this, however, was the reason why I paused upon reading Cymro's challenge to the readers of the *English Mechanics*. No. It was rather a tumbling of years that checked my thoughts. Indeed, Cymro's question was 51 years old when I first encountered it, and now question 128 has passed into its 93rd year. How old is this question? I asked myself. I felt sure that Cymro was not the first person to pose this challenge, and so it was that my task, this task, unfolded. Who was it that initially formulated the question, and why? For indeed, Cymro's question is a highly contrived one, and it is even predicated upon an impossibility; no such Earth-crossing tunnel could ever be engineered — or can it? Talking as a physicist or even as an engineer might, one could possibly argue that a student would be better challenged, in a pedagogical sense, to build up a list of the many reasons why the question is so poorly founded. For all this, the question has a certain sense of wonderment about it; dig a tunnel straight through the Earth — imagine that.

The first solution to Cymro's problem was provided by an *English Mechanics* reader simply identified as C.P. No. 1, on March 21st — just two weeks after the problem was printed. A second answer to the problem was given by S. A. Swann in the March 28th issue of the magazine. Swann's solution is presented in the Appendix, while the motion of a particle falling through an Earth-tunnel is illustrated in Figure 3.1. The answer to Cymro's problem, as pointed out by Mr. Swann, and as

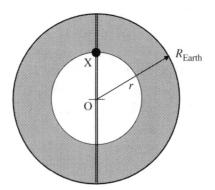

Figure 3.1.   The geometry of Cymro's problem.

---

[2] Pun intended ☺.

revealed in Figure 3.1, is based upon an application of Newton's shell theory, and he writes, "Consider the particle in a position X, distance OX = r from the center O. With center O describe a sphere to pass through X. The Earth has thus been divided into (1) a concentric shell, inclosing (2) a concentric sphere." From here, Swann notes, "It is known from the elementary theory of attraction [Newton's shell theory] that (1) exerts no attraction at a point inside it, and that (2) attracts an external point as if its whole mass were concentrated at the center O." The other important part of Cymro's question is that it states that the Earth is to be taken as a "uniform sphere." This simply means that the mass per unit volume inside of the Earth is to be taken as being constant — in other words, it has a constant density ρ. The Earth most certainly does not have a constant-density interior, as we shall see later, but if this assumption is not made then there is no straightforward solution to Cymro's problem — indeed, it becomes a decidedly complex mathematical problem. The importance of assuming a constant density is that the mass of material M interior to a sphere of radius R can be conveniently expressed as $M = (4\pi R^3/3)\rho$, where the term in the brackets is the volume of a sphere of radius R. Where this becomes important in solving Cymro's problem is that we now have the understanding that as the radius becomes smaller, not only does the sphere become smaller, but so too does the mass of material contained within the sphere. Additionally, and this is really the more important point, the acceleration that a sphere of radius R and mass M will induce upon a small particle of mass m located upon its surface can be calculated as: $a = F_{gravity}/m$, where $F_{gravity} = -GMm/R^2$, where the minus sign indicates that the force is directed towards the Earth's center. For a uniform sphere, however, we have just argued that the mass M contained within the sphere is $M = (4\pi R^3/3)\rho$, and upon substituting for this term in our equation, we find that $a = -(4\pi/3)G\rho R$. What this result tells us is that the acceleration experienced by the small particle is directly related to the radius of the sphere R upon which it is imagined to sit at each instant, and accordingly, as the sphere becomes smaller, so too, in direct proportion with the radius, does the acceleration.

With these details in place, a word answer to Cymro's problem may now be articulated. When OX corresponds to the Earth's radius, R_{Earth}, the small particle, assumed to be initially at rest, will experience the full gravitational acceleration due to the entire Earth — this is the Earth's surface gravity, which can be measured experimentally and comes out to be 9.81 m/s^2. Accordingly, the particle once let fall, will descend down the tunnel, towards the center of the Earth[3] with a continually increasing speed. As the particle moves down the tunnel, however, at any given position OX = r, where r is smaller than R_{Earth}, the gravitational attraction resulting from the entire shell of material above r, that is from r to R_{Earth} (the shaded region in the figure), has no effect upon the particle's motion — this is Newton's shell theory at work. The acceleration that the particle does experience, however, will be that due to the sphere of radius r, and accordingly the acceleration must be smaller than that which applied at the Earth's surface, because r is smaller than R_{Earth}. In this manner, as the particle approaches the Earth's center, its speed increases, but its instantaneous acceleration decreases — the latter effect coming about because r is getting smaller and smaller. At the Earth's center OX = r = 0, the speed will then be at its maximum value, while the acceleration will be at its smallest possible value — zero in fact. Even though the acceleration has dropped to zero at the Earth's center, the particle keeps on moving along the tunnel since it has some finite velocity. Of course, having passed through the center of the Earth, the particle is on its way back out towards the surface, and its motion will begin to slow down because now the acceleration is in a different direction — always towards the center. Now the mass of material interior to its location is increasing (because r is increasing) and accordingly there is a larger and larger gravitational force acting to slow the particle down. Eventually, at the far end of the tunnel to which it was initially released, the particle will come to a complete standstill, with zero velocity, but be subject to maximum acceleration (the Earth's surface gravity) that will once more direct the motion of the particle back down the tunnel towards the Earth's center. Here we have the oscillatory part of the answer emerging. The particle will accordingly fall back down the tunnel and the whole motion, back and forth across the Earth's center, will be repeated *ad infinitum*. The motion just described is illustrated in

---

[3] Here it is assumed that the tunnel has been excavated along the Earth's spin axis, all the way from the North Pole to the South Pole — but see Collignon's Slant, Chapter 17, later on.

Figure 3.2, which is based upon the detailed mathematical solution given in the Appendix.

The top panel in Figure 3.2 shows the position r of a small particle as it moves through the tunnel against the time since it was released into the entrance — the variation is given by the relation: $r/R = \cos(\omega t)$, where R is the Earth's radius, t is the time and $\omega = (g/R)^{\frac{1}{2}}$. The bottom panel in

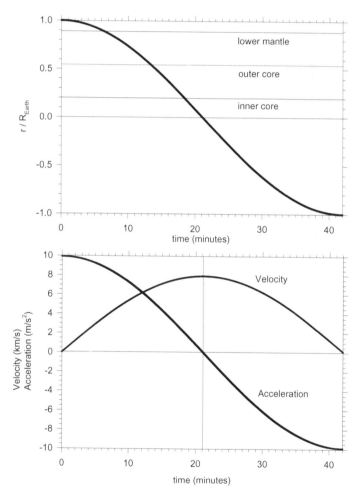

Figure 3.2. The solution to Cymro's problem, showing (*top*) the variation of the distance traveled (as a fraction of the Earth's radius), (*bottom*) velocity and acceleration as a function of time. The horizontal lines in the top panel indicate the various boundaries between the surface, the mantle and the Earth's core.

Figure 3.2 shows the variation in the velocity (initially zero) and the acceleration (initially a maximum of 9.81 m/s²) against the time since the particle was let fall into the tunnel. When the particle is at the Earth's center, the velocity, as described earlier, is at its maximum value, some 7.9 km/s as it turns out, while the acceleration is zero. The detailed calculation (see the Appendix) indicates that the Earth-crossing time for the tunnel particle is $t_{tunnel} = \pi(R/g)^{\frac{1}{2}} = 42.2$ minutes. That is, it takes the particle some 21.1 minutes to reach the Earth's center, and a further 21.1 minutes to reach the far end of the tunnel. Comparing the tunnel crossing time with the analogous hoop crossing time we discover that $t_{hoop} = (2/\pi)t_{tunnel}$. This result indicates that the Earth-crossing time in the analogous hoop model is about 26.9 minutes; some 1.6 times faster than that of the actual Earth-tunnel crossing time. The reason for this difference relates to the fact that in the hoop model the acceleration is constant and acts continuously, whereas in the Earth-tunnel problem the acceleration continuously changes and is reduced to zero at Earth's center.

We now have a descriptive answer to Cymro's problem — we have used a little mathematics to justify our words, and we have relied entirely upon Newton's description of how forces, specifically the gravitational force, acting upon specific bodies produces changes in their position, velocity and acceleration. We have also made use of Newton's shell theorem to justify our answer. Our descriptive answer (as well as the mathematical solution given in the Appendix) is, in modern terms, as complete as it can be. At different times in history, however, Cymro's basic question, what happens to a particle falling into an Earth tunnel, might have been phrased in different ways and decidedly different conclusions about its motion would have been reached. These different, historical responses will be considered in later chapters. For the present, however, let us take a thought-experiment journey through an Earth tunnel, and use the results shown in Figure 3.2 as our guide to developments.

If I should fall through this Earth (as a crepuscular body under Cymro's conditions), what would I see and experience? Here, if ever there was one, is a question derived from a dark and disturbing dream. The solid Earth; the constant Earth; the Earth beneath my feet: how could it not support the weight of my body? Let the nightmare continue — the very ground swells and my feet slip in a sickening frictionless slide through the topsoil. Within half a second I am six-feet deep — a fleeting occupant of the land of the dead and our long-buried ancestors. But there is no eternal rest as my body picks up speed and plummets downward. After 5 seconds I am already 120 meters below the Earth's surface, and by 10 seconds' time I am nearly half a kilometer into my journey, but at this stage I have hardly penetrated the Earth's crust, and the rocks around me (assuming that I can see them) are the same as those which are found at the surface. My speed of fall continues to grow, and by 30 seconds I have travelled 4.4 kilometers from the Earth's surface and my speed is close to 300 meters per second or 1,080 kilometers per hour. I am now several hundred meters deeper into the Earth than the bite of the Mponeng Gold Mine in South Africa. This mine cuts to the deepest levels yet achieved by the human desire for wealth and trinketry, and in the mine's lowest corridors the temperature of the rock exceeds 65 degrees Centigrade. Hell is seemingly approaching, but remarkably not only do thousands of people work under these oppressively hot conditions, there are the Mponeng ghosts — illegal workers who actually live in the mine shaft for months on end and eke out a living by secretly refining stolen ore and ferrying the extracted gold to the surface. After one minute of fall time I am just over 17.5 kilometers below ground and traveling at a speed of some 600 meters per second. I have now passed beyond the greatest depths to which any human being has descended or human drilling operation has ever reached. The deepest borehole to be cut into the Earth's crust is that by the Kola Superdeep Borehole in the Murmansk Oblast region of Russia. Before the Kola project closed operations in 1987, it managed to reach a sample depth of 12.26 km. The borehole diameter of the Kola shaft was just 23 cm across, but through this narrow opening core samples were routinely extracted and remarkably bacterial life within the rocks was found to thrive to a depth of at least 6 km below the surface. I am now, however, the only living entity (at least in this thought experiment) for the next 6,365 km to the Earth's center. At this stage I have not yet passed through the Earth's crust — that thin shell of rocks that forms the entire arena of active geology and which supports the domain of shifting plate tectonics.

For all this, after about 80 seconds from the start of my fall, I crash through the boundary, at a depth of some 30 kilometers, between the crust and the mantle; my speed of descent is now about 1 kilometer per second. Between the crust and the upper mantle resides a thin transition zone called the Mohorovičić discontinuity. Named after the Croatian geologist Andrija Mohorovičić, this boundary was identified in 1909 on the basis of seismic studies. Indeed, the discontinuity induces a sharp change in the propagation speed of earthquake waves, and it is additionally associated with a distinct change in the types of rocks that the earthquake waves must move through. Above the discontinuity the rocks are predominantly silicates, similar to those found at the Earth's surface, while below the discontinuity they are basalts — rocks relatively depleted in silica, but rich in other minerals such as pyroxene and olivine. For the next five minutes, my speed still increasing, I descend through upper mantle rocks (see Figure 3.3); the temperature is now of order a thousand degrees Centigrade and the surrounding rocks are near molten and malleable. After some 13.5 minutes from the start of my fall, I am nearly 3,000 km below the Earth's surface, half way to the center, and my speed is now about 6.5 kilometers per second. At this time I am close to the

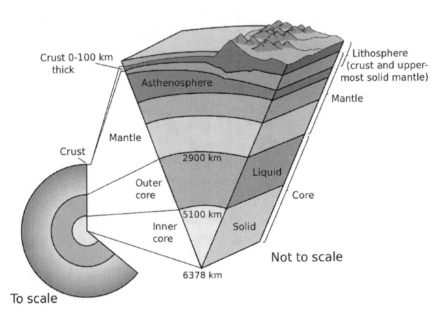

Figure 3.3. Cut-away diagram showing the Earth's interior structure and identifying the major mantle-core dichotomy. Image courtesy of USGS.

Earth's outer core, where the rock changes to a liquid quagmire of nickel-iron. That the Earth actually has a core, distinct in composition and structure to the mantle, was first suggested, on the basis of seismogram data, by Irish-born geologist Richard Oldham in 1906. Rushing ever downwards, however, 19.5 minutes into my fall, at a speed of some 7.8 kilometers per second, my surroundings once more change in dramatic fashion and I am now falling through a mass of crystalline nickel-iron. The temperature of my surroundings is some 7,000 degrees and the pressure is a crushing 3.6 million atmospheres — hell, by any other name. For the next minute and half, I rush through the Earth's solid inner core, finally reaching the center of the Earth after 21 minutes of fall time. At the center my speed is just a little shy of 8 kilometers per second. Having attained maximum speed at the Earth's center, I am now heading, feet first, towards the surface again. After another 21 minutes I will emerge, rising feet first, like an inverted demon, from the Earth's surface soil.

Chapter 4

The Heat of Ages

If our eye could penetrate the Earth and see its interior from pole to pole, from where we stand to the antipodes, we would glimpse with horror a mass terrifyingly riddled with fissures and caverns.

Thomas Burnet, 1694.

Ateshgah, the Temple of Fire, located at Baku in Azerbaijan on the western side of the Caspian Sea, is a place of great historic veneration, having been used at various times as a religious temple by the Hindu and Zoroastrian faithful. The temple is a humble, but elegantly constructed stone structure, and it contains a natural marvel — an eternal flame.[1] Indeed, the entire region of the Absheron peninsula is famous for a landscape that oozes oil; where lakes of boiling mud bubble, churn and steam; and where flames lick upwards from the ground. It is through these naturally occurring phenomena that Azerbaijan gains its literal name: The Land of Flames. If ever the traveler needed proof that below the nurturing crust of the Earth sits a burning inferno, then they need travel no further than the flame-licked landscape of Azerbaijan. Even without the eternal flame of Ateshgah, however, the idea that the temperature increases with depth underground is long-founded. The rise in temperature, at least in the thin veneer part of the crust that has been explored by humans, amounts to about 25 degrees Centigrade per

---

[1] Indicating that there truly is an end to all things, the eternal flame flickered out in 1969, and the temple's fire is now fed by a commercial gas pipe connection.

kilometer of depth underground — just 4 kilometers down, the temperature is already at the boiling point of water. Picking up on this very point, British physicist William Thomson (later Lord Kelvin of Largs) observed in 1862 that "the fact that temperature increases with depth implies a continued loss of heat from the interior by conduction outwards or into the upper crust." The key questions, however, are where does all this heat come from and does the temperature continue to rise and rise with depth, ultimately reaching the point at which rock will melt and ooze. Volcanoes immediately tell us that subterranean temperatures can and do exceed that at which rock will melt and flow, but the broader question about the Earth's internal heat supply is less easily answered.

Throughout all human history, natural disasters predicated on volcanic paroxysms have been recorded — they are recorded in the ash-embalmed dead of Pompeii, a citizenry brought down by the eruption of Mount Vesuvius in A.D. 79, and they are recorded in the staggering death tolls resulting from the eruptions of Mount Tambora in 1815, Mount Krakatoa in 1883, Mount Pelee in 1902, Mount St. Helens in 1980, and in countless other outbursts throughout time. The stricken are invariably caught unawares, ignoring the pre-eruption rumblings and tympanic paradiddles, as if they are just the by-product of wrestling giants at play underground. Even experts can get it wrong. Tragically, Stanley Williams, Professor of Volcanology at Arizona State University, and companions were caught in an unexpected eruption while exploring Mount Galeras in Colombia in 1993. In an interview with the *BBC Science Focus* magazine, Williams recalls that "in an instant, Galeras went from doing nothing to screaming, roaring, shaking the ground, and throwing rocks in all directions …. My friends in the crater were vaporized, shredded into tiny particles at a thousand degrees Centigrade." Nine people died in this one unexpected eruption. The power and wrath of the Earth, when in a boisterous mood, are truly staggering and typically those who experienced it only have a few more moments to live.

For all this, volcanologists continue to risk life and limb in order to better understand those magmatic vents that puncture Earth's surface with such violence. Saint Januarius[2] (3rd century A.D.) is the official

---

[2] The faithful still gather three times a year in Naples Cathedral to witness the liquefaction of what is claimed to be a sample of St. Januarius's blood preserved in a sealed glass ampoule.

patron Saint of Naples and volcanic eruptions, although a better (none-martyred) candidate might be the German-born Jesuit scholar Athanasius Kircher. Born in 1602 Kircher grew up to become an inquisitive soul, ever determined to investigate the world and document all its creatures and phenomena. He wrote extensively on many topics, but it was his first book (to be discussed later), *Ars Magnesia*, published in 1631, that propelled him to fame, and ultimately brought him to work and study at the Collegio Romano — the Jesuit school established by St. Ignatius of Loyola in 1551. Kircher's investigations into the workings of the Earth are summarized in his book *Mundus Subterraneus* published in 1664. Kircher's book is a walloper, consisting of nearly 1,000 pages of text spread over 2 books divided into 12 sections, and by way of introducing its contents he explains that his plan is to set out "before the eyes of the curious reader all that is rare, exotic and portentous contained in the fecund womb of Nature."

We learn about Kircher's researches into the structure of the Earth in the English translation of *Mundus Subterraneus* in the wonderfully entitled, *The Volcano's [sic] or, Burning and Fire-Vomiting Mountains, Famous in the World* (published in 1669). The key elements of Kircher's depiction are essentially laid out in the first few lines of Chapter 1, which begins, "That there are subterraneous conservatories, and treasures of fire (even as well, as there are of water and air, etc.) and vast abysses, and bottomless gulphs in the bowels and very entrails of the Earth, stored within, no sober philosopher can deny." The picture (Figure 4.1) that Kircher develops is that of an Earth beset with a spider's web of fire channels and caverns, all interconnected and feeding into the volcanoes located at Earth's surface. Kircher has been described as being the last polymath, and he developed many of his ideas along the lines of accepted ancient Greek wisdom and contemporary religious thinking. He did, however, set out to investigate first-hand the phenomena of Mount Vesuvius in 1638. Kircher writes beautifully and expressively, indeed, in a manner that evokes the very thrill of adventure: [sic] "Having a very earnest desire, a long time, to understand the miracles of subterraneous nature .... I undertook a voyage into Sicily and Malta.... I ascended the Mountain at midnight, through difficult, rough, uneven, and steep passages. At whose *crator* or mouth, when I had arrived, I saw what is horrible to express; I saw it all over of a light fire, with an horrible combustion, and stench of

Figure 4.1. Athanasius Kircher's *Pyrophylaciorum* showing the Earth's central fiery core and an inner system of interconnected channels and fire conservatories.

Sulphur and burning bitumen. Here forthwith being astonished at the unusual sight of the thing; me-thoughts I beheld the habitation of Hell; wherein nothing else seemed to be much wanting, besides the horrid fantasms and apparitions of devils." While Kircher, one foot firmly placed in the medieval past, probably felt he was truly gazing into the fire caverns of hell venting atop of Vesuvius, he went on to develop his ideas concerning the Earth in tones similar to (but with distinct differences nonetheless) that in the modern-day Gaia concept. To Kircher the Earth was alive and the fire caverns were part of a life-sustaining purification process. A giant whirlpool at the North Pole, Kircher argued, fed icy water into the Earth's interior where it trickled and percolated through and around the warming fire channels, eventually emerging, warmed and purified at the South Pole. This breathing and purification cycle of the Earth kept the seas from freezing over and stopped them from turning rancid and stagnant.

With one foot in the past and one foot in the future, Kircher straddles the boundary between ancient and modern thinking. On one page we find him developing new ideas that resonate with modern observations, and on other he can be found writing about dragons and underworld giants, with additional discussion on astrology, alchemy and the spontaneous generation of insects from dung. For all this, Kircher is attempting to develop, at least in part, a working model of the Earth — an Earth that has an active, hot interior and an Earth containing natural features and mechanisms that can explain the rising and lowering of tides, earthquakes, and volcanoes. The Earth's central fire may well have been a vision of hell to Kircher, but it was additionally the life-sustaining heat pump of a living and breathing world. Given that the Earth is hot within its interior, the question immediately arises as to how long this heat supply can last and continue to power geological activity. The presence of volcanoes on Earth indicates that there is still plenty of internal heat to melt rock and drive tectonic motion, but modern-day images of Mars also reveal volcanic domes and caldera, but they are all now long dead and inactive.

In general, the amount of heat contained in an object will vary according to its volume, while the rate at which it loses energy into space, through heat radiation, will depend upon its surface area. The cooling time t_{cool} of a spherical object of radius R will vary, therefore, according to its radius: $t_{cool} \sim (4\pi R^3/3)/(4\pi R^2) \sim R$, and this tells us that small objects must cool off more rapidly than larger objects.[3] Taking the best (but limited) information available at the time (circa 1860) William Thomson (Lord Kelvin) set out to determine the age of the Earth by solving the equation describing heat diffusion. He argued that the initial temperature of the fully molten Earth could not have been more than several thousands of degrees, which is the characteristic melting temperature of rocks and metals. Accordingly, he found that for the Earth to attain its current surface temperature, based upon various estimated rates for surface heat loss, it would require a time interval of some 20 to 400 million years. This age estimate for the Earth, derived as Kelvin put it "according to elementary physical principles," was embarrassingly small

---

[3] This result begins to explain why Mars is no longer volcanically active — since it is half the size of the Earth, so for a given heat content it will cool off twice as fast. Current survey techniques indicate that Mars has not been volcanically active for the past half billion years.

for the geologists and the evolutionary biologists, and a largely belligerent battle of egos ensued. The then predominant uniformitarean school of geologists argued on the principle that all of Earth's topological features must have been formed through processes that were slow, miniscule, additive and continuously in operation, and that this required many hundreds of millions, to even billions of years' worth of acuminated time in order to account for Earth's features. Likewise the evolutionary biologists argued that many hundreds of millions of years were required for natural selection to bring about the diversity of species evident in the biosphere, and to explain the fossil record. Kelvin, however, stuck to his physics guns, and even set out to argue that the Sun was sensibly hotter mere millions of years ago. Writing in *Macmillan's Magazine* for March 5th, 1862, Kelvin argued that it was "most probable that the Sun has not illuminated the Earth for 100,000,000 years, and almost certain that he has not done so for 500,000,000 years." The two sides could not agree on the answer concerning the age of the Earth, but interestingly it was Kelvin, in his *Macmillan's Magazine* article that hinted at the eventual solution. Not only did Kelvin estimate the age of the Sun, and hence the Earth by association, in his magazine article, he also set out to determine for how much longer the Sun might shine, writing "the inhabitants of the Earth cannot continue to enjoy the light and heat of their life for many million years longer." To this statement, however, he added an all-important boiler-plate condition, "unless sources now unknown to us are prepared in the great storehouse of creation." By this latter reasoning, Kelvin was essentially saying that his calculations could be wrong, that is, giving too short a lifetime, if the Sun is able to generate internal energy by means other than that associated with slow gravitational collapse. And this situation is exactly that which has played out; the Sun generates its internal energy not through shrinkage, but through fusion reactions converting hydrogen into helium within its central core. The physics of atomic fusion, however, and especially the details of how such physics might operate within stars were not to be understood until the late 1930s, long after Kelvin's death. It transpires, as we shall see below, that this same boiler-plate condition provides an answer to why it is that the Earth is some 10 times older than Kelvin's maximum deduced age for the Sun.

Mary Dickson and Mario Fanelli of the *Instituto di Geioscienze e Georisorse*, in Pisa, Italy have estimated that the total heat content of the Earth amounts to some 10^{31} joules.[4] Other geophysical measurements indicate that the Earth radiates heat from its surface regions at a rate of about 47 terawatts — that is 4.7×10^{13} joules per second — corresponding to a surface flux of about 0.1 joules' worth of energy per square meter per second. The gradual loss of heat energy at the Earth's surface dictates a slow cooling off of the Earth over time. An estimate of the Earth's future cooling time, that is, the time t_{cool} to radiate (given no further additions) all of its thermal energy, follows from the division of the heat content by the surface heat loss: accordingly t_{cool} is about 7 billion years. At the end of these 7 billion years all of the Earth's internal heat energy will have been radiated away, and all plate tectonic motion and volcanic eruptions will grind to a halt — indeed, the Earth will become geologically dead. This future moribund state may come about, but there is now good evidence from computer models of star evolution that the Earth will, in fact, be destroyed and consumed in the fiery atmosphere of the bloated, red-giant Sun some 5 billion years from the present time.

In the modern era it is reasoned that there are two primary sources for the Earth's internal heat: one is primordial and the other is actively on-going, beneath our feet, to this very day (and into the future beyond). The primordial heat content is reflected in the formation mechanism for the Earth — a process of accretion and growth, through collisions with massive asteroid, even Mars-sized bodies, which operated in the newly forming solar system some 4.5 billion years ago. The impact energy of the accretion process was transferred into internal heat, and the proto-Earth is envisioned as being molten throughout its interior — a glowing ball of incandescent molten rock and metals. Over a time span of several hundred million years the Earth grew larger and larger, accreting more and more mass, until emerging at the end of the growth phase with its

---

[4] The joule, named after James Prescott Joule, is usually defined in terms of the energy transferred to an object when a force of 1 newton acts upon it over a distance of one meter. Alternatively, a calorie is defined as being the energy required to raise the temperature of one gram of water by one degree Celsius, and 1 calorie is equivalent to 4.184 joules.

present mass and an accompanying Moon.[5] Indeed, the Moon formation event probably left the Earth in a fully molten state with a global magma ocean. After this Moon-forming cataclysm, the Earth slowly began to cool off, developing a fragmentary solid crust, and acquiring a differentiated interior in which the denser elements (iron, for example) began to sink towards the core with the lighter elements (i.e., rock) precipitating towards the outer layers. The current division of the Earth's interior into a rocky crust, mantle and nickel-iron core (recall Figure 3.3) is a direct result of its primordial molten state, and about half of the 47 terawatts of energy that is currently leaking out from the Earth's interior is thought to be residual heat from its tumultuous formation. If only half of the Earth's surface heat flux is primordial, where then is the rest of the energy coming from? To answer this, we have to look at the instabilities of atoms and the process of radiogenic heating.

That some naturally forming elements can cause covered and sealed photographic plates to darken was first noted by Henri Becquerel in 1896. The mysterious agents responsible for this darkening were called Becquerel Rays, and it was initially thought that they might have properties in common with those of X-ray radiation — the latter being discovered by Wilhelm Röntgen in 1895. This connection, however, turned out to be unfounded: X-rays have a commonality in fact with light, being a form of electromagnetic radiation, while Becquerel rays are actual subatomic particles. One of the most active producers of Becquerel rays is Uranium, and experiments on this and similar such substance, conducted by Becquerel, Ernest Rutherford, and Pierre and Marie Curie, revealed that a complex chain of events was actually taking place. The generation of the rays was accompanied by the gradual transmutation of one element into another, and this transmutation took place at a steady exponential rate. Additionally, it appeared that there were two types of *ray*: these are composed of the so-called alpha and beta particles, as first described by Rutherford in 1899. Later research showed that the alpha particles were helium atom nuclei, composed of two protons and two neutrons; the beta particles, which are much more penetrating than

---

[5] The Moon formation event is generally thought to have taken place some 80 to 100 million years into the Earth accretion process.

alpha particles, turned out to be electrons — the electron being first discovered by Joseph Thomson in 1897. What Becquerel had discovered in 1896 was not a new form of electromagnetic radiation, but the *disjecta membra*[6] associated with radioactive decay. With the added realization that radioactive decay can act as a potential heat source within terrestrial rocks, it was realized by Rutherford in 1904 that Lord Kelvin's argument concerning the cooling age of the Earth could be modified. Indeed, Rutherford solved the problem by evoking the same boiler plate that Kelvin had mentioned with respect to his limiting 500 million years' age for the Sun. The point, as Rutherford explained, was that Kelvin's age limit was based entirely upon the diffusion rate of primordial heat; given, however, that there was an active internal energy sources within the Earth's interior (the decay of radioactive elements), the Earth's age could easily be pushed back to times encompassing many billions of years. Rutherford's experience during his 1904 lecture has gone down in the annals of science history, and he recorded that "To my relief Kelvin [who was in the audience] fell asleep, but as I came to the important part, I saw the old bird sit up, open an eye and cock a baleful glance at me! Then a sudden inspiration came, and I said, 'Lord Kelvin had limited the age of the Earth, provided no new source of heat was discovered. That prophetic utterance refers to what we are now considering tonight, radium.' Behold! The old boy beamed at me." That the Earth and solar system must be truly ancient, with a formation time set some 4.5 billion years into the past, was first experimentally verified by American chemist Claire Patterson in 1956. Patterson's remarkable measurement was based upon a study of the lead isotopes produced by the decay of radioactive elements found within a sample of Canyon Diablo iron meteorite. Since meteorites are derived from asteroids, and asteroids are in turn the left-over building blocks of the planets, so the planets, Earth included, must be of a comparable age to that deduced for the asteroids.

Numerous radionuclides have been identified and amongst the most important, with respect to heating the Earth's interior, are uranium-238, thorium-232 and potassium-40. That radioactive decay can cause a heating effect is a simple consequence of the conservation of energy, since

---

[6] Latin for "scattered fragments."

the alpha and beta particles are physically ejected from the nucleus during a decay event, and the kinetic energy of their motion is transformed into the thermal motion (that is heat) of the surrounding material. The amount of energy that can be generated through the decay of a specific radionuclide depends in part upon its so-called half-life, which is literally the time required for half of the parent element to decay into the stable daughter element. For example, uranium-238 eventually decays to lead-206 after the ejection of 8 alpha particles and 6 beta particles, but this multi-stage process runs on a half-life timescale of $t_{1/2} = 4.468$ billion years. Accordingly, if $N(0)$ is the initial amount of uranium-238 contained within the young Earth, then the amount of uranium-238 that will exist at some later time t, will be $N(t) = N(0)\exp(-0.693t/t_{1/2})$. Given such a long half-life, uranium-238, although a far-from-abundant natural element, can play its part in the slow cooking of the Earth's interior. Indeed, the slow decay of uranium-238 is such that it takes about 340 billion kilograms of mantle rock in order to generate one joule of energy per second through radioactive decay. Present-day geothermal models place the radioactive materials predominantly in the Earth's lithosphere and mantle, and suggest that within this region some 16 terawatts' worth of energy is generated by the decay of uranium-238 and thorium-232, with a further contribution of 4 terawatts coming from the decay of potassium-40. This gives a grand total of about 20 terawatts — an energy generation rate that is comparable to that from primordial heat loss from Earth's core.

Beneath our feet, driven by primordial heat and atomic decay, the warmed mantle rocks slowly churn, driving through their undulations the drift of continental plates and the venting of new magma along volcanic mid-ocean ridges. The continental plates drift and meld at speeds amounting to perhaps a few centimeters per year, and slowly, over eons, these motions change the very appearance of the globe — political lines on the map be damned. And, while the heating and the indomitable heaving of the lithosphere take place in regions well beyond our direct gaze and exploration, there are, nonetheless, fleet-footed particles that tell us about the deep, inner-Earth heat budget — these are the ghost-like geoneutrinos. Neutrinos are produced as a natural by-product of the β particle decay process, but unlike the alpha and beta particles, neutrinos have virtually no mass and they travel at close to the speed of light. The existence of such strange particles was inferred by Italian physicist Wolfgang Pauli in the 1930s, at which time it was realized that something was carrying energy in the decay of a neutron (n) into a proton (p^+) and an electron (e^-). Initially it was argued that since a neutron has no associated charge, when it decayed into a proton, which has a positive charge, an electron (with its associated negative charge) must also be produced — in this way $n \rightarrow p^+ + e^-$, where the $\rightarrow$ indicates the decay taking place, and where the minus-sign superscript attached to the electron symbol e^- indicates its negative charge characteristic (the + superscript for the positively charged proton p is usually not shown, but is included here for clarity). While this decay of the neutron accounts for charge conservation, it does not, experiments show, account for energy conservation. That is, the energy associated with the neutron before the decay is greater than that associated with the proton and electron after the decay. Somewhere, it seemed, energy was not being accounted for correctly, and it was this realization that led Pauli to suggest a third neutral particle was involved in the decay, and that this unseen particle carried away just the right amount of energy to ensure that energy was conserved before and after the decay. Pauli called the new particle a neutrino, which is from Italian for little neutral particle. With Pauli's addition, the decay of the neutron is now written as: $n \rightarrow p^+ + e^- + \nu$, where ν is the neutrino (technically it is an anti-electron neutrino[7]). In terms of neutrino sources, the Sun provides the dominant flux at the Earth's orbit — indeed, something like 61 billion solar neutrinos are passing through every square centimeter of your body, every second, right now.

To say neutrinos don't interact easily with matter is rather an understatement. The so-called cross-section area of interaction for a neutrino is estimated to be a minuscule 10^{-47} m^2. Casting this estimate in other terms, if a neutrino is imagined to be moving through a thin column of pure water, then the column will need to be at least 18 light-years long before even one interaction event is likely to take place — that is a column of water stretching over 4 times further into space than the distance

---

[7] There are actually 6 neutrino flavors — the electron, tauon and muon neutrinos and their antineutrino counterparts.

to Proxima Centauri, the nearest star to the Sun. In spite of their remarkably reluctance to interact with matter, the first direct detections of neutrinos were made in the mid-1950s. In order to capture such an elusive prey, experimental physicists have to present the incoming neutrino flux with as many targets for interaction as possible. In the case of neutrinos from the Sun, the first successful detector was a giant cylinder containing 615,000 kg of perchloroethylene, which provided the incoming neutrinos some 2×10^{30} clorine-37 atoms to interact with. Even with this vast number of potential target atoms, an exposure time of between 50 to 100 days was required before a measurable detector response could be generated [7]. In terms of detecting neutrinos produced in the decay of uranium and thorium within Earth's mantle

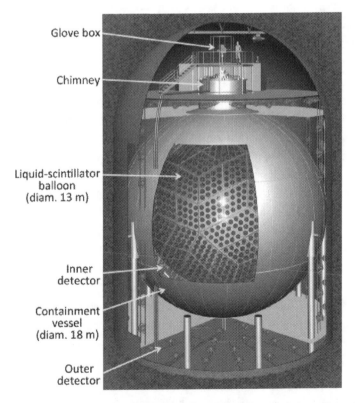

Glove box
Chimney
Liquid-scintillator balloon (diam. 13 m)
Inner detector
Containment vessel (diam. 18 m)
Outer detector

Figure 4.2. Schematic outline of KamLAND. The human figure at the top of the diagram (by the calibration "glove box" system) gives some idea of the experiment's scale.

(so-called geoneutrinos), the first detections were made in 2005 with the Kamioka Liquid Scintillator Antineutrino Detector (KamLAND[8]) in Japan (Figure 4.2). It is estimated that along with the dominant solar neutrino flux, there are an additional 6 million geoneutrinos passing through each square centimeter of our bodies every second at the Earth's surface at the present time. It is remarkable to think that we spend our entire lives literally swimming through a diaphanous sea of ghostly neutrinos, and we are completely unaware of it.

---

[8] This remarkable detector consists of a 13-meter diameter nylon balloon containing some 1000 kg of mineral oil, benzene and fluorescent chemicals. The detector fluid contains of order 6×10^{31} target protons for the electron antineutrinos to interact with, via what is known as the inverse beta decay. The detector records a single geoneutrino event every three to five days.

Chapter 5

Terricola's Questions

Mathematics is not a deductive science — that's a cliché. When you try to prove a theorem, you don't just list the hypotheses, and then start to reason. What you do is trial and error, experimentation, guesswork.

Paul Halmos (mathematician)

Serendipity will eventually take one to magical places; luck, tenacity and the shear shucking of calumny will, sooner or later, land one in the right place, on the right page, at the right time. It is this very possibility that ultimately keeps the historian going; it is this belief that keeps them digging through dust-encrusted tomes, and hunting down those more than obscure references. The origin of Cymro's question was not, it turns out, buried so deep within the historic literature that it had passed into impenetrable shadow — but it did require a good many hours of frustrated searching in order to find. Such is the life of the academic and historian. While I did end up making a number of journeys down interesting, but ultimately fruitless, *cul-de-sacs*, the answer concerning the origin of Cymro's problem did eventually emerge — and I, for one, was surprised by the answer.

As with any search through historical material it is better to start from a known recent position and then work further backwards. It is the case that Cymro's question can be found in almost any present-day physics and/or calculus text. Phrased one way or another, it will be there, rest assured. But, of course, what about textbooks which were published further back in time? Fortunately there are not that many books of relevance to this topic that were available more than a hundred years ago.

There are classic texts, however, and some of these are still in print to this very day or are at least generally held within university library collections. In answering Cymro's problem, our anonymous writer CP No. 1 suggests that he, that is Cymro, might want to consult "Tait and Steele's *Dynamics of a Particle*" and this seemed like a promising place for me to begin my search as well.

Peter Guthrie Tait was a famous and well-respected mathematician of his time. He graduated senior wrangler[1] in the fiendishly difficult University of Cambridge Mathematical Tripos examinations for 1852 (William Steele was second — *Proxime Accessit* — wrangler in that same year), and over a highly productive life he worked on aspects of thermodynamics with Lord Kelvin, and he was one of the founders of topological knot theory, and is particularly remembered for his work in graph theory and the four-color map problem. Steele's career was sadly cut dramatically short, however, and he died of consumption, at age 23 years in 1855. In spite of Steele's pre-mature death, Tait included his name in the authorship of his first publication, *A Treatise on the Dynamics of a Particle* (Cambridge University Press, 1856). The text is a classic of its time; sparingly written and overly dense with equations — it is not a text for the mathematical faint of heart. For all this, the wisdom, skill and methodology displayed in Tait and Steele is outstanding and it certainly holds its own against any modern-day text on dynamics. Sure enough, however, on consulting Tait and Steele, I found in Chapter III the Earth-tunnel problem. The problem is introduced with no specific fanfare and no accreditation is given to its origin, and while Tait comments in the preface to the text that many of the examples in the book were taken from Cambridge University exam papers, it is not clear if the Earth-tunnel question was, as such, an exam question. For all this, the fact that Cymro's question was included in Tait and Steele, its origin can be pushed back to at least 1856, some 68 years prior to its *English Mechanics* airing. A slightly later text than Tait and Steele is that by Edward Routh, *A Treatise on the Dynamics of a Particle, with Numerous Examples*, this confusingly similar titled text

---

[1] The Mathematical Tripos was taken by those students at Cambridge University studying for the degree of Bachelor of Arts. After working through a grueling set of examination papers, the students were ranked in order of merit, with those in the first class being designated wranglers. A history of the Tripos examinations is given by W. W. Rouse Ball in his *Mathematical Recreations and Essays* (first published in 1892).

being first published in 1898. Routh was a famed mathematics coach at Cambridge and his text contains numerous college exam and Tripos questions, and while it does contain Cymro's problem as an exercise it is not labeled as being a specific exam question.

From Tait and Steele, the next text I surveyed was that by John Henry Pratt: *The Mathematical Principles of Mechanical Philosophy, and their Application to the Theory of Universal Gravitation* (Deighton Press, 1841). Pratt studied mathematics at Cambridge (graduating as third wrangler in 1833), was a member of the select group of *Cambridge Apostles*, and after joining the British East India Company in 1838 became Archdeacon of Calcutta in 1850. Pratt is particularly remembered for his analysis of the gravitational anomaly, caused by the Himalayan Mountains, on the vertical suspension of a plumb line. His text is much more formidable than that by Tait and Steele, and was clearly not written for the mathematical neophyte. For all this, Cymro's problem is to be found (somewhat hidden) in Pratt's text. Specifically, Pratt poses the question, "to find the attraction of a homogeneous body differing but little from a sphere in form upon an internal particle." While the answer to this question is developed in terms of the parameter $e = (1 - \text{polar radius/equatorial radius})$, it is evident that when e is set to zero (corresponding to a perfect sphere) the problem is exactly that as posed by Cymro (see also Chapter 22). Once again Pratt offers no specific reason or origin comment with respect to the question he asks and then answers, but its inclusion and corollary push the Earth-tunnel question back a few more years to at least 1841.

The next text I consulted was William Whewell's *An Elementary Treatise on Mechanics* (Cambridge University Press, 1819). Whewell was an extraordinary polymath who wrote extensively on many topics from poetry to mathematics, philosophy to astronomy and on the possibility of extraterrestrial life. His text was "designed for the use of students in the university" and while wordier than the text by Tait and Steele it is a comprehensive mathematical introduction to statics and dynamics. Cymro's problem is not posed in Whewell's text since he only considered motion under a constant gravitational force — but he does pose in Chapter III, problem 1, a near-surface equivalent to Cymro's problem with respect to dropping a stone into a deep well (recall Chapter 2, and see the Appendix). At this stage I had only pushed the origin of Cymro's question back to 1841, and was beginning to run out of additional texts to review.

Continuing along the path that Cymro's question might have constituted a university exam or indeed, a publicly posed problem, the most obvious book to look through next was the comprehensive text compiled by Thomas Leybourn. Leybourn was a mathematics teacher at the Royal Military College in London for 37 years, but he is largely remembered today for his editorial work on numerous mathematical journals and diaries. Of specific interest here was his role as editor and organizer, from 1799 to 1835, of *The Mathematical Repository*. The *Repository* was solely concerned with the publication of mathematical questions as posed and answered by its readers. The questions were an eclectic combination of geometry, pure mathematics, dynamics, number theory and combinatorics. The readership of the *Repository* was presumably equally as eclectic and many of its most accomplished contributors were highly qualified mathematicians working under various pseudonyms — we find answers and questions from such readers as, spitfly, Merones Minor, Astronomicus, Tyro Philomatheticus, and Scoticus. Once again, and as with the readers of *The English Mechanics*, only a few of the pseudonyms can be attached to recognized figures. A search through all volumes of the *Repository*, a non-trivial exercise I might add, yielded no question similar to Cymro's problem. Usefully, however, in what is called *The New Series of the Mathematical Repository*, published between 1806 and 1830 and bound in 5 volumes, are reproduced, without answers, the entire set of Cambridge University Senate House Examination papers from 1811 to 1829, and these did yield some intersecting tidbits. By modern standards these examination papers are an extravaganza of mixed-discipline questions that take the examinee from problems relating to number theory, to problems on differential calculus, to physics, to astronomy and geometry. The examination questions reproduced in Leybourn's *Repository* would probably strike fear into many a present-day physics or mathematics student, and, for example, in 1830, the following wonderfully obtuse question is found,[2] "A body descends down the convex side

---

[2] Here we recall the words of Sir Frederick Pollock, senior wrangler in 1806, "a Cambridge education has for its object to make good members of society — not to extend science and make profound mathematicians." Pollock became a well-known lawyer and scholar of his time, writing numerous books on jurisprudence and contract law. He taught at the University of Oxford, was a member of the Society of Authors, a member of the Privy Council and in later life was elected Treasurer of Lincoln's Inn.

of a logarithmic curve placed with its asymptote parallel to the horizon, find where it leaves the curve." Amongst the Tripos question reproduced in the *Repository*, I found no exact match to Cymro's problem. What I did find, however, were the loop and shell versions of Cymro's problem, as described earlier in Chapter 3. These two questions were set by Richard Gwatkin, a renowned private tutor at Cambridge, in the 1819 examinations. I came across two questions that are close in form to that of Cymro's problem — one set in the 1823 examinations and one in the 1828 examinations. The 1823 question reads, "shew that the velocity acquired by a body in falling from infinity to the Earth's core, is to the velocity of a secondary at the Earth's surface as $\sqrt{3} : 1$" [8]. This question has superficial links to the Earth-tunneling problem in that it requires the derivation of an expression for the gravitational potential inside of a constant density sphere. The 1828 Tripos question reads, "A body falls towards a center of force which varies as some power of the distance, determine the cases in which we can integrate so as to find the time of descent." This question contains (but does not specifically ask for) the answer to Cymro's problem — the specific case being when the central force F varies as $F = kr$, where k is a constant and r is the distance from the center (recall Chapter 2). The more general answer to the Tripos question, however, reaches back to an analysis, and famous thesis, on particle dynamics, published in 1736, by Swiss mathematician Leonhard Euler (as we shall see in Chapter 16).

Having found no hint of Cymro's problem within Leybourn's *Repository*, I turned to his other well-known compilation, *The Mathematical Questions Proposed in the Ladies' Diary, 1704–1816* (published in four volumes in 1817), and here I hit pay dirt. *The Ladies' Diary: or Woman's Almanack* was founded by English mathematician John Tipper in 1704. The *Diary* featured material relating to astronomical phenomena (sunset/rise, Moon phases and so on) along with recipes, medical advice and short stories. There was additionally a scientific and mathematics section in which the editor and *Diary* readers posed various short problems to be solved; there was also a more substantive, that is complex, annual Prize Question. Leybourn clearly recognized the pedagogical value and quality of the mathematical questions being posed in the *Diary*, and his 1817 compilation was a detailed list of all the questions submitted, along with their solutions. The first mathematics question to be presented in Leybourn's collection reads, "In how long a time would a million of millions of money be in counting, supposing one hundred pounds to be counted every minute without intermission, and the year to consist of 365 days, 5 hours, 45 minutes."[3] Similar such questions would appear in most issues of *The Diary*, but as time progressed the questions became more and more challenging. Question 238, posed by Mr. Peter Kay in 1743, reads, for example, "to find in what arc of a circle a pendulum must vibrate, so that the time of one whole descent shall be equal to the time in which a heavy body would fall along the chord of the same arc." Question 497, by Mr. Stephen Ogle, posed in 1762 reads, "Required to draw a right line through the focus of any given Apollonian parabola, so as to divide the area thereof into two such parts as shall obtain a given ratio." The reader invariably had a wide choice of questions to attempt, with topics ranging from dynamics, statics, geometry, algebra, trigonometry, series summation, astronomy and surveying.

Indeed, the *Diary* questions compiled by Leybourn are a vast extravaganza of mathematical delights and brain numbing nightmares. Eventually, however, alighting on question 784, as posed in 1781, my heart rate took a leap. The question reads, "if a ball be let fall from the surface down a perforation made diametrically through the Earth, it is required to find its velocity and time of falling to the center, and to any given point, with other circumstances of its motion, abstracted from the effect of the Earth's rotation, and on the supposition that the Earth is a homogeneous sphere 8,000 miles diameter." Here is Cymro's problem, now pushed back by a further 60 years from Pratt's *Mathematical Principles*.

Question 784 was posed under the *nom de plume* of Terricola, which translates from the Latin to *a dweller upon the Earth*. The correct solution to question 784 was provided by Mr. Robert Phillips of St. Agnes in Cornwall, England — the solver's name in this case presumably being actual rather than made up. It is not clear who Robert Phillips was, but he is found providing answers to various mathematical questions set not just in the *Diary*, but in the *Gentleman's Monthly Intelligencer* and *The Town and Country Magazine*. He was clearly an accomplished mathematician, and his name, for example, is found amongst the subscribers list to Henry Clarke's, *A Dissertation on the Summation of Infinite Converging Series with Algebraic*

---

[3] The answer is 19013 years, 144 days, 5 hours, 55 minutes.

Divisors (published in 1879).[4] Phillips provides a compact solution to Terricola's question, correctly deducing the velocity at the center of the Earth to be $V_{center} = (Rg)^{\frac{1}{2}}$ and that the travel time to the center will be $t_{center} = (\pi/2)(R/g)^{\frac{1}{2}} = 21.1$ minutes (indeed, just as we saw in Chapter 3 and as shown in CP. No. 1's solution in the Appendix). Interestingly, Phillips adds a corollary to his answer arguing on purely physical grounds that the motion of the ball will be periodic, oscillating "forward and backward continually." It is not clear why Terricola did not ask for this additional condition to be proved in his question. Indeed, even though Phillips uses fluxions (differential equations) to get his answer, he does not find the general solution, which as seen in the Appendix provides for an answer in terms of a time-varying cosine term. It is for this reason that Phillips's argument that the ball will oscillate up and down the tunnel is based upon the idea of symmetry, rather than any specific mathematical solution supplied.

I have not been able to find any older versions of Cymro's problem in the literature, and it seems that its origins trace no further back than 1781 — although much older variants in which the Earth is not assumed to be homogeneous do exist, as we shall see. Taking Terricola as the originator of Cymro's problem, therefore, the question now becomes: who was Terricola?

Up to this point in our narrative it has been difficult, if not impossible, to place names and faces to the various *nom de plumes* and pseudonyms encountered. A few such connections can be made, however, and Terricola turns out to be British mathematician and 5th Astronomer Royal the reverend doctor Nevil Maskelyne (Figure 5.1). In 1781 Maskelyne would have been 49 years old, and 16 years into his 46-year residency as Astronomer Royal. His career at the time of setting the *Diary* question was well established and he was one of the preeminent astronomers of his time. Maskelyne's reputation had been firmly established in the 1770s when he had overseen a series of observations at Schiehallion, a mountain located in Perthshire, Scotland, in order to determine the Earth's density. These observations determined the apparent difference in latitude, at two stations on either side of Schiehallion, as measured with a plumb-line and as traced out by direct triangulation. It was found

---

[4] Clarke was a frequent correspondent and problem solver to *The Ladies' Diary* and he was also a close friend of Charles Hutton, the *Diary's* editor from 1774 to 1818.

Figure 5.1. The reverend doctor Nevil Maskelyne (1732–1811).

that these two measures gave slightly different results, and the difference was attributed to the gravitational attraction of Schiehallion. By estimating the mass of Schiehallion[5] it was then possible to deduce an average density for the entire Earth. With the density of the Earth determined, it was then possible, for the first time, to calculate the mass of the Earth directly — as we shall see in Chapter 10, the actual size of the Earth had been well-determined by the mid-18th century. At the time of writing his question for the *Diary*, Maskelyne would have been heavily involved with the editing and production of the so-called *Requisite Tables*[6] (published in May 1781) to accompany the *Nautical Almanac and Astronomical Ephemeris*, as well as supervising the observational work at the Greenwich Observatory, and organizing the Board of Longitude chronometer trials. It is remarkable, therefore, to find Maskelyne having time for any other

---

[5] Schiehallion was specifically chosen because it is a relatively isolated, triangular-prism-shaped mountain.

[6] The *Requisite Tables* set out the procedure by which longitude at sea can be determined through observations of the Moon's position.

activities and especially mathematical recreations. Certainly Maskelyne was an accomplished mathematician, graduating from Trinity College, Cambridge as 7th wrangler in 1754, and it is clear that he was both well equipped and well able to develop some fiendishly complicated mathematical problems. Indeed, under the pseudonym of Terricola, Maskelyne wrote two annual prize questions for the *Diary* — the first in 1795 and the second in 1801. It is worth seeing these two prize questions since they reveal a deep sense of subtlety and abstraction. The 1795 question reads, "Suppose the whole terraqueous globe taken as a sphere, should be instantaneously turned into a uniform elastic aeriform fluid, whose particles repel one another with a force which is to that which those of air repel one another, as the density of one to the density of the other, it will expand itself either to a finite or infinite extent, still preserving the form of sphere. It is required to determine the force of gravitation tending towards the center, and also the density, at any given distance from the center; supposing the mean density of the Earth to be 3,825 times that of air at the surface of the Earth." The 1801 question reads, "A quantity of matter being given, it is proposed to determine the figure of a solid of rotation made up of it, which shall have the greatest possible attraction on a point at its surface." The answers to these questions need not directly detain us here (they do involve some considerable unpacking — see the Appendix); what is more important is the fact they illustrate some remarkable abstract thinking (thought experiments no less) that relate, admittedly in a superficial way, to real-world geophysical problems. The two prize questions as well as the Earth-tunnel problem (question 784) have a sense of logical connectedness, as thought experiments, about the Earth's physical structure and the dynamical balance between gravity, always acting to make an object as small as possible,[7] and the internal pressure forces resisting the effects of gravity. Here is a very different Maskelyne than the normal histories reveal: it is a man contemplating entirely abstract physical questions, and not the stalwart "Seaman's Astronomer" intent upon making navigation at sea safe [9].

---

[7] Indeed, if no internal pressure forces opposed gravity then the Earth would rapidly collapse, in modern parlance, into a black hole. The gravitational collapse time for the Earth, in the situation that all opposing forces suddenly vanish, is just 15 minutes.

It is not known to what end Maskelyne developed his particular questions for the *Diary*, although they do have a resonance with his 1770's work concerning the measurements made at Schiehallion and in the determination of the Earth's mass [10]. Indeed, the Editor of the *Ladies' Diary* at the time when Maskelyne submitted his questions was his long-time friend Charles Hutton. It was, in fact, Charles Hutton that did most of the detailed (and arduous) calculation work relating to the Schiehallion project. Hutton was for many years professor of mathematics at the Royal Military Academy in Woolwich, and was elected a fellow of the Royal Society of London in 1774. It is not beyond the realms of possibility (although there is no direct evidence to support the idea) that both Maskelyne and Hutton worked on the development of Terricola's questions; perhaps even discussing and preparing the answers to the problems to be solved. We see some evidence of this in Hutton's writings in the early 1800s. In his *Recreations in Mathematics and Natural Philosophy*, published in 1803, Hutton includes the questions, "if we suppose a hole bored to the center of the Earth, how long time would a heavy ball require to reach the center, neglecting the resistance of air?" At first the answer is given under the condition of uniform acceleration, which as we shall see later is Galileo's solution to the Earth-tunnel problem, but he adds a note to this answer stating that "it is more probable that a body, proceeding along the radius of the Earth, would lose its gravity, as it approached the center [...] and it can be demonstrated, supposing the density of the Earth be uniform [...] that the gravity would decrease in the same proportion as the distance from the center." It seems odd, at first glance, for Hutton to give what is essentially an incorrect answer to the problem and then add a note to the fact that the answer is wrong. The probable reason for this is that Hutton's *Recreations* text was based in large part upon a translation of Jacques Ozanam's 1708 book, *Recreations Mathematical and Physical*. Ozanam includes a question concerning the fall time of a stone dropped into "a well as deep as the center of the Earth," but completely fails to give a correct answer — indeed, Ozanam argues that the fall time will be "almost 3 hours." Strangely, however, Ozanam provides a correct argument with respect to the oscillation of the stone if the "well continued to the antipodes" — essentially, it would seem, Ozanam provides Galileo's thought-experiment answer to the Earth-tunnel problem (see Chapter 12), but fails to grasp the mathematical aspects of the physical conditions of free fall under constant acceleration conditions.[8] Hutton, in his translation, appears to have corrected Ozanam's mathematical error in the constant accelera-tion fall time for the stone, and then attempted to improve upon the physical actuality of the question. Indeed, in a later text published in 1811, *A Course of Mathematics in three volumes for the use of Academics as well as private tutors*, Hutton entirely ignores Galileo's constant-acceleration Earth-crossing thought experiment, but includes Terricola's question, lifting it *holus bolus*, and answering it in the same manner as outlined by Robert Phillips in the 1781 *Ladies' Diary*.

Cymro's problem, as classically laid out, and as given by Terricola, is predicated on the notion that the tunnel cuts across the Earth's diameter, and therefore the stone, particle or corpuscle dropped into the tunnel passes through Earth's core. Cymro, however, does not specify in his ques-tion that the tunnel has to pass through the Earth's center — it can be any straight line path between two points on the Earth's surface. The first text that I encountered allowing for this specific freedom was that by Sidney Luxton Loney, professor of mathematics at the Royal Holloway College, Surrey, in his book, *An Elementary Treatise on the Dynamics of a Particle and of Rigid Bodies*, published in 1909: "assuming that the Earth attracts points inside it with a force which varies as the distance from the center [this is equivalent to saying that the Earth is a homogeneous sphere], shew that, if a straight frictionless airless tunnel be made from one point on the Earth's surface to any other, a train will traverse the tunnel in slightly less than three quarters of an hour." In this case the tunnel need not cut through the Earth's core, but it turns out that the end result of the motion and the travel time across the tunnel are identical to the diame-ter-crossing tunnel. While Looney provides no background to his Earth-tunnel question, we shall see in Chapter 17 that it builds upon an analysis of the Earth-tunnel problem, published in 1883, by French civil engineer Édouard Collignon. It is also a problem that caught the whimsical eye of Oxford mathematician Charles Dodgson, as we shall see in Chapter 18.

---

[8] Under the constraint of constant acceleration, the descent time to the Earth's center will be ½ t_{hoop} = 13.3 minutes — recall Chapter 3.

While the inspiration behind Maskelyne's penning of the Earth-tunnel problem for the *Ladies' Diary* is not known, one possible answer may reside in his place of work — the Royal Observatory at Greenwich. Founded in 1675 by King Charles II, the Greenwich Observatory is one of the early instances of a government-funded scientific research program. In characteristic "keeping up with the Joneses" style, the Observatory at Greenwich was the British response to the construction and opening of the Paris Observatory, under the patronage of the Sun King (Louis XIV of France), in 1671. The Greenwich Observatory, and the instruments that it would contain, were largely left to the designs of the Royal Society, Robert Hooke and Christopher Wren, although the choice of the 1st Astronomer Royal, John Flamsteed, proved to be controversial. Flamsteed was a perfectionist and seeming able to pick a fight with, and carry a grudge against, almost anyone. For all this, his observations eventually resulted in the publication of a great catalogue, *Historia Coelestis Britannica*, containing the positions of 2,935 stars — the accuracy of these star positions far out-ranking those of any other catalogue that had hitherto been published. By the time that Maskelyne became 5th Astronomer Royal, the Greenwich Observatory had been in operation for some 90 years, and its main research efforts were directed towards perfecting the Moon position tables, vital for the determination of longitude, and the publication of *The Nautical Almanac and Astronomical Ephemeris*. These programs required intensive amounts of calculation and complex reduction of astronomical observations, and it was Maskelyne's job to make sure that it was all done correctly and in a timely fashion. For all this, however, he did find time for his mathematical puzzles and for the composition of an occasional poem. Perhaps, therefore, while strolling amidst the grounds of Greenwich Observatory, musing upon a poetical composition perhaps [9], his thoughts might have turned to the mystery of the location of Flamsteed's Well, which, by that time, had long been filled in, although it was, in fact, the superstructure of one of the first astronomical instruments to be used at the Observatory [11].

Chapter 6

Flamsteed's Well

Through the gorge that gives the stars at noon-day clear.

Rudyard Kipling, *The Song of the Banjo*, 1894

Look up. Look way up. What do you see? Well, you see the point directly overhead. To the astronomer the view overhead, the zenith, has special properties. Firstly, the zenithal view provides the astronomer with the shortest possible view through Earth's atmosphere, and accordingly it is the least affected direction with respect to atmospheric disturbances. Light from a star in the observer's zenith travels directly through the atmosphere, and the effects of atmospheric refraction, which otherwise bends the path of starlight, is accordingly near zero. The view in the zenith is literally the straight goods. Astronomers have long exploited this zenithal condition, and in the process have produced some of the strangest of telescopes. These same telescopes, however, have also revealed fundamentally important astronomical results. But first, before we discuss the zenith telescope, we have an old-wives tale to deal with.

The story begins with Aristotle — who had an answer for everything even if he had never observed the phenomenon under question. Within his discussion on the *Generation of Animals*, written circa 350. B.C., Aristotle explains to the reader that "people in pits and wells sometimes see the stars." Pliny the Elder in his *Natural History*, written in 79 A.D., alludes to the same possibility, and somehow or other this complete myth has managed to perpetuate itself, at least in popular culture, to this very day. Alexander Humbolt in Volume 3 of his *Cosmos* (the entire 5-volume collection being published between 1845 and 1862) picks up the story, and records that during his time as an inspector of mines he was always on the lookout for stars being visible through deep shafts. He never saw any stars, and nor did he ever encounter any miners who had seen stars through a mine shaft either. Astronomer John Herschel, in his *Outlines of Astronomy* (first published in 1849) records the myth, but adds that no chimney sweep that he had ever quizzed upon the possibility had ever seen stars during the day through a chimney. Jules Verne, in contrast, writing in his *Journey to the Center of the Earth* (first published in 1864) described with full literary gusto, but great scientific inaccuracy, the apparent daytime visibility of β-Ursae Minoris from the bottom of a 1-mile deep volcano shaft. Likewise, in George Elliot's *Adam Bede* (first published in 1859) we find Mrs. Irvine alluding to the hapless man "who stands in a well and sees nothing but the stars." The anonymous "M.D" further recounts in the pages of the *Astronomical Register*, for February 1865 that "I have more than once, when voyaging down a creek in Equatorial America, seen stars by day. Towards evening the lofty and leafy forest excludes horizontal rays, and the only light that reaches the voyager comes from the chasm in the foliage where the trees fail quite to arch over the creek. In these strips of blue a twinkling star may often be seen by the traveler." Hugo Gernsback in his *Scientific Adventures of Baron Munchausen* (serialized between 1915 and 1917) also alludes to his hero being able to see stars through a 2,164-mile long Moon tunnel (see Chapter 9 later). Indeed, it would require the remarkable powers of Baron Munchausen to see such a phenomenon, and not only do the actual observations negate Aristotle's claim, so too do the detailed scientific studies [12].

While there is no optical advantage in looking up a coal-mine shaft, or a tall chimney, in order to see the stars, this is not to say that stars can never be seen with the human eye during daylight hours. The restriction, however, is to the brightest stars (Sirius, Canopus and Aldeberan) and the planets — these may be seen, if one knows where to look, when the Sun is close to the horizon. Why then, one might ask, did the 1st Astronomer Royal, John Flamsteed record in his diary that on July 30th, 1679, that he had descended to the bottom of a 100-foot well that had been dug into the hill at Greenwich in order to observe the heavens? The answer, it turns out, was an attempt to continue an experiment first attempted by Robert Hooke — the well was the terrestrial tube of a great telescope

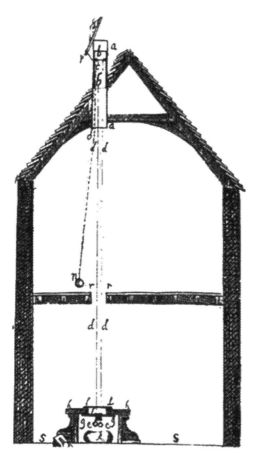

Figure 6.1. Hooke's zenith telescope. The lens is located at *b* in the tube projecting through the roof. The observer's head rests on the pillow *k* on the floor *s-s*.

pointing straight up into the sky. Indeed, the idea behind the well telescope was to measure stellar parallax [13]. Hooke had invented and then installed a zenith-pointing telescope in his home in 1669 (Figure 6.1) and had attempted to measure the six-monthly shift in the zenith angle of the somewhat inconspicuous star γ-Dracons (Eltanin),[1] which importantly passes directly overhead for observers in London. On the basis that such

---

[1] Eltanin is the 3rd brightest star in the constellation of Draco. It is located about 150 light years away and is some 2 times more massive, 50 times larger and 470 times more luminous than the Sun.

a parallax could be measured, the physical distance (in units of the Earth's orbit about the Sun) to Eltanin could be calculated. This would then reveal the distance to at least one star (other than the Sun) and give some measure to the extent of the stellar realm. It was a fundamental science project, but unfortunately while the theory was entirely correct, the technology of the 17th century was not up to the challenge. Hooke's experiment with his zenith telescope failed when the lens he was using fell from its mount; for Flamsteed it was the cold and damp, as well as faulty optics that compromised the observations. As far as the official record shows no more than a handful of observations were ever made from the well at Greenwich.

It is not known if the telescope well was dug by design at Greenwich or whether it was adapted from an existing structure. From the diagram produced in Flamsteed's *Historia Coelestis Britannica* (published in 1712), the well was well-constructed, being some 7-feet in diameter, brick-lined for most of its length and incorporating some 150 steps in order to descend the 100-feet to its terminal cavern. Drop a stone into Flamsteed's well and it will be some 4½ seconds before the annoyed yell of the astronomer would come back to one's ear (see Figure A1 in the Appendix). The lens for the telescope was obtained by Sir Jonas Moore from the French optician Pierre Borel and was being tested by Flamsteed in early 1677; the lens, which still survives to this day,[2] is some 9¾ inches in diameter and has a focal length 87½ feet. The well itself no longer exists, the shaft being filled in some time prior to 1737. During various work projects carried out at the Observatory, in 1790 and 1881, the brick work at the top of the well has been *rediscovered*, but it was not until the late 1960s that several detailed surveys and archeological excavations were performed to confirm the well's location, which is now marked by a plaque in the Greenwich grounds.

While the well telescope did not prove successful with respect to measuring the parallax of Eltanin, Flamsteed did perform a detailed set of observations, with a small fixed-position telescope, relating to the transit times of Sirius (the brightest star in the sky — next to the Sun that is).

---

[2] The lens was returned to the safekeeping of the Royal Society of London by James Hodgson (Flamsteed's long-time astronomy assistant) in 1737. The lens was later presented, in 1932, to the Science Museum, South Kensington.

His aim here was to confirm the steady rotation rate of the Earth. Indeed, through such transit observations gauged according to the exquisitely made, year-going Tompian clocks that had been installed at the Observatory, Flamsteed concluded in 1678 that there was no measurable variation in the Earth's rotation rate. Even though Robert Hooke had overseen the design and construction of the astronomical instruments initially employed at the Greenwich Observatory, he does not appear to have been greatly interested in actually using the zenith telescope well. This apathy was presumably the result of a dislike for Flamsteed and the fact that he had designs on another longer-focal-length zenith telescope. This latter zenith telescope, in fact, is the massive, fluted Doric column of the Monument to the Great Fire of London (Figure 6.2). The great fire occurred in September 1666, and Robert Hooke, along with Christopher Wren, was preeminent in directing the rebuilding of inner London. The design for the monument (most probably by Hooke) was approved by City Council in 1671, but it was another 6 years (1677) before the great column was finally completed [14]. The interior of the monument's column is hollow and it was intended that the full 202 feet of its length would be used to house a zenith-pointing telescope. It is not clear if this final adaptation ever came about, and there are no records of observations having been conducted from the monument's location. The great power of the zenith telescope, however, was to be proven in 1725 when James Bradley (the future 3rd Astronomer Royal) along with Samuel Molyneux successfully employed a 12½-foot-long device at Wanstead to detect stellar aberration.[3] Bradley later used the same instrument (between 1727 and 1747) to show that the direction of the Earth's spin axis nutates, that is nodes up and down, with an 18.6-year period as it moves around its 26,000-year precession circle. The zenith sector so successfully used by Bradley was eventually moved to the Greenwich Observatory in 1749; it was moved again in 1837 and installed (for a two-year stint) at the Cape of Good Hope Observatory in Southern Africa.[4]

---

[3] Along with the then newly discovered finite speed of light (by Ole Rømer in 1676) the six-monthly variation in the aberration offset provided the first clear experimental indication that the Earth was in motion about the Sun.

[4] The telescope was acquired by the National Maritime Museum in 1955 and it is currently on display (in its original 1749 position) at the Greenwich Observatory.

Figure 6.2. The Monument to the Great Fire of London, designed by Robert Hooke (probably in collaboration with Christopher Wren). The column was so constructed that it could act as a 202-foot long zenith telescope. The depiction of the monument shown here is by Sutton Nicholls circa 1753.

The zenith sector and telescope at Greenwich was never destined to measure stellar parallax; that was a set of measurements eventually to be made by a triumvirate of astronomers circa 1838, when the six-monthly modulation in the sky positions of 61-Cygni, Vega and α-Centauri were first revealed.

Was the rumor of Flamsteed's lost well and zenith-pointing telescope Maskelyne's inspiration for the 1781 Earth-tunneling question? The answer to this we may never know for certain. When Maskelyne first became Astronomer Royal, Flamsteed's well would have been out of commission for at least 28 years. For all this, however, memories at grand institutions tend to be long, and it remains a distinct possibility that the deep well with telescopic eyes was Maskelyne's inspiration.

Chapter 7

Airy Underground

I am a jovial collier lad and blithe as blithe can be,
For let the times be good or bad they're all the same to me.
'Tis little of the world I know and care less for its ways
For where the Dog Star never glows, I wear away my days

Down in a coal mine, J. B. Geoghegan (1872)

While you can't see stars from the bottom of a mine shaft, you can attempt to weigh the Earth, and this is exactly what George Biddell Airy (the future 7th Astronomer Royal) and his friend William Whewell set out to do in 1826. In this case it was not the use of the mine shaft as a zenith-pointing telescope that was being exploited, but rather it was a means by which Newton's shell theory could be put to work. Recall from Chapter 3 that Newton's shell theorem dictates that the gravitational acceleration that a person at the bottom of a mine will experience is directly related to the mass of material beneath their feet, and not to the shell of material above them. This small difference in the gravitational mass of the Earth, that at the top of the mine shaft and that at the bottom, can be measured with a simple pendulum. At issue is the measure of the number of swings the pendulum will make over a given time interval, since a pendulum's oscillation period T is dependent upon the length of the pendulum wire l and the local gravitational attraction g, with $T = 2\pi(l/g)^{\frac{1}{2}}$. The period of a pendulum will change, therefore, if one changes the length of the support wire, or if one changes the value of the local gravitational attraction, or both. With the experiment conducted by Airy and Whewell the idea was to keep the length of the support wire fixed, so that any variation in the pendulum's oscillation period at the top of the mine shaft and at the bottom would be due to the change in the local gravitational attraction.

In their 1826 experiment, Airy and Whewell chose to use the Dolcoath mine in Cornwall [15]. This was one of the deepest mines in England at that time, and it allowed for pendulum variations to be measured over a vertical distance of some 600 meters top to bottom. In Airy's later words, "the labor of the experiment was excessively great (principally because the ascent and decent are effected entirely by ladders)." The pendulum experiment conducted by Airy and Whewell is straightforward in principle, although very complex in practice. The idea is to set up two synchronized clocks and two associated free pendulums at the top and bottom of the mine shaft. Then, once all is assembled, the experiment involves counting the number of swings that each pendulum makes over a given time interval of many hours. If there is any difference in the local gravitational acceleration, then the number of swings completed by each pendulum will vary in proportion to the ratio of the gravitational acceleration at the top and bottom of the mine shaft. This difference, with knowledge of the local density of rocks and surroundings, can then be used to determine the Earth's density, and this in turn can be combined with a measure of the Earth's radius (see Chapter 10) to determine the Earth's mass. The tricky part of the experiment is not in the counting of the pendulum swings, although that is arduous enough, but in compensating for the effects such as variation in the length of the pendulum wire due to temperature differences at the top and bottom of the mine, and variations in atmospheric pressure and humidity affecting the operation of the clocks. Nonetheless, Airy and Whewell persevered against the conditions only to be denied a satisfactory result by the destruction of their apparatus by a fire in the mine. Our intrepid pair tried the experiment again in 1828, and were once more defeated, at the last minute, by a rock fall and the flooding of the lower mine shaft.

Here the situation rested for some 26 years, after which time Airy (now Astronomer Royal) along with a whole team of assistants set up shop at the Harton Coal Mine in South Shields [15]. There, over a three-week period, Airy and team set about making good on a series of pendulum experiments at the top and the bottom of the 1,260-feet

(384 meters) deep mine. This time luck smiled upon Airy; not only were there no mine calamities, they had mechanical cages in which to transport the equipment, and the observers additionally had "a galvanic [signal] system" which allowed coded messages and time markers to be sent between the two locations. Following several months of calibration and comparison of the results, it was announced in December 1854 that the pendulum at the top of the Harton mine completed 2.5 fewer oscillations per day than the pendulum at the bottom, implying that the gravitational attraction at the bottom of the mine is, in fact, higher than at the top. This result, unfortunately, is the exact opposite to expectation [16], since the simplest application of Newton's shell theory would predict that the top pendulum should complete more swings per day than the bottom one. The Harton Colliery result actually informs us that the Earth is not a homogeneous sphere (see Chapter 18). Compensating for the density of the coal bed strata, however, Airy was eventually able to deduce that the average density of the Earth was some 6.6 times that of water. This is not an unreasonable result, since it lies almost mid-way between the density of a typical Earth-mantle rock (about 3,300 kg/m^3) and that of nickel-iron (about 8,000 kg/m^3), but it is an overestimate by some 20 percent when compared to the accepted value of 5,514 kg/m^3 in the present day.

While the mine-based pendulum experiment works in theory, it is a very difficult experiment to perfect, and it was the laboratory-based torsion balance technique that ultimately provided a definitive value for the Earth's average density. Often called the Cavendish experiment, after British natural philosopher Henry Cavendish, the experiment was initially suggested by John Mitchell in the early 1780s. Cavendish perfected the experiment and obtained the first results between 1797 and 1798, deducing a mean density for the Earth of 5,448 kg/m^3 — within 1 percent of the modern-day value. While Cavendish technically found a better value for the mean density of the Earth (by modern reckoning) than that by Airy some 56 years later, repeat torsion balance experiments, by other researchers, showed a large range of variation.[1] Indeed, it was not until

---

[1] Once again it is the controlling of environmental conditions that is important — this and the fact that very small deflections of a pendulum bar need to be measured.

the opening up of the 20th century before consistent results for the Earth's mean density were determined [17].

The pendulum is one of the simplest devices in the physicists' arsenal of experimental tools, and yet it is also one of the most fundamental [18]. Exquisitely precise measurements of gravitational variations can be made with the pendulum and geologists have long exploited the power of the pendulum to map out subsurface topology, find oil fields, and to measure the propagation and effects of earthquake waves. Not only this, however, the pendulum can provide us with several thought-experiment conundrums. What, for example, is the oscillation period of a pendulum of infinite length oscillating close to the Earth's surface? Now, here is a physically impossible pendulum, but it is presented within a question that has, in fact, a well-defined answer. If we look one more time at the formula used previously for the oscillation period of a pendulum, we have $T = 2\pi(l/g)^{\frac{1}{2}}$, so one might argue that as the length l of the pendulum becomes longer and longer, the oscillation period will become longer and longer, with the ultimate result that as $l \rightarrow \infty$, $T \rightarrow \infty$. This, however, is the wrong answer. Figure 7.1 illustrates the situation envisioned in our question. The infinitely long (assumed massless!) pendulum wire stretches to the right (forever), and the Earth is taken as a finite sphere of radius R and surface gravity g. Since the pendulum mass is taken to be moving close to the Earth's surface the force acting upon it will be $F = mg$, where m is the mass of the pendulum bob. We can now think of resolving the force that acts upon the pendulum bob into the x-direction and the y-direction. The oscillation is taken to be in the y-direction and at any general location (as shown in the diagram) $F_y = F\sin(\phi)$. The force acting

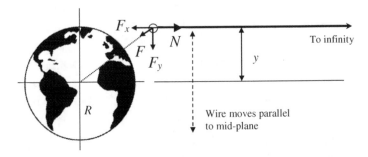

Figure 7.1.   The infinite-length, near-Earth pendulum.

in the x-direction is balanced by the tension N in the pendulum wire, and this is always constant. Accordingly, the equation of motion for the pendulum bob will be (via Newton's second law): $F = ma_y = -F_y = -mg(y/R)$, where a_y is the acceleration in the y-direction, R is the Earth's radius, and where the substitution $\sin(\phi) = y/R$ has been made. The equation of motion now reads as $a_y = -Ky$, where $K = (g/R)$ is a constant term, and this equation describes what is known as simple harmonic motion (see the Appendix). Furthermore the period of oscillation for a simple harmonic oscillator is: $T = 2\pi(R/g)^{\frac{1}{2}}$.

Remarkably, there is a limit to how slow a pendulum (not already at rest) can actually move when located at or near to the Earth's surface, and the period of oscillation is determined according to the size of the Earth. Even if some super pendulum is constructed with a wire longer than Earth's radius it will still only oscillate with a period set by $2\pi(R/g)^{\frac{1}{2}}$. More than this, of course, if we check back to Chapter 3, we also see that the infinite pendulum has an oscillation period identical to that of a particle dropped into an Earth tunnel. This is a wonderful convergence of two impossible experiments that can only be carried out in the thought-experiment universe. That the finite Earth-tunneling problem has both contact and resonance with an infinite length pendulum (suspended from nowhere?) is an entirely satisfying result. It is also a result that makes sense, since the arc of an infinite-length pendulum is, of course, a straight line. It is also a somewhat surprising result since the infinite-length pendulum assumes that the gravitational attraction is constant, while in the Earth-tunneling situation the gravitational acceleration is changing from a maximum at Earth's surface to being zero at Earth's center. There is yet more in this result to surprise us, however, since it also turns out that the oscillation period of the infinite-length pendulum is identical to that of the minimum period of an Earth-orbiting satellite.

Given a satellite that just skims around the surface of the Earth (see Chapter 14 later), and ignoring any interaction with the atmosphere, Kepler's third law [19] provides us with a relationship between the satellite's orbital period P and its orbital radius R (which is taken to be the Earth's radius), such that $P^2 = (4\pi^2/GM)R^3$, where G is the gravitational constant and M is the mass of the Earth (here, as is only possible in thought-experiment world, the mass of the satellite is completely ignored). Close to the Earth's surface, however, the gravitational acceleration is constant and is given by $g = GM/R^2$. A little rearranging then reveals that $P = 2\pi(R/g)^{\frac{1}{2}}$. This result further informs us about the speed of the satellite V, since the time to complete one orbit around the Earth's surface is simply the Earth's circumference $2\pi R$ divided by speed V. We recover, therefore, from our two relationships for the period that $V = (Rg)^{\frac{1}{2}}$. This speed, upon substituting numbers, corresponds to some 7.9 km/s and it is exactly the same as the maximum speed that the Earth-tunneling particle will have as it crosses the Earth's center.

Chapter 8

A Mind's Eye View

Thursday March 29 [...] we had fuel to make two cups of tea apiece and bare food for two days on the 20th [...] I do not think we can hope for any better things now. We shall stick it out to the end, but we are getting weaker, of course, and the end cannot be far [...] it seems a pity, but I do not think I can write more.

Last entry from the journal of R. F. Scott

The thought of actual adventure, by which I mean with boots and tents, and ropes and danger, totally appalls me. If Thoreau was an *outdoorsman*, ensconced amid the minimalism and simplicity surrounding Walden Pond, then I am very definitely an *indoorsman*. I need regular meals, cups of tea, and a comfortable bed at the end of the day. Camping holidays are one thing, and long walks and bicycle rides are certainly there to be fully enjoyed, but the discomfort of true adventure is for the pleasure of others to experience. I am not suited mentally, equipped physically, nor built appropriately for hard days against savage cold, wind-driven rain and adversity. If at first you can't climb the mountain, go home and enjoy a good book — that's my motto. Indeed, if there is any adventure that I am best suited for, *indoorsman* that I am, it is that of mental journeying. In sympathy with R. L. Stevenson:

"There, in the night, where none can spy,
All in my hunter's camp I lie,
And play at books that I have read
Till it is time to go to bed"

Adventure of the mind, adventures from the page, these I can follow safely, and travel in the safe footsteps of those who have suffered the ordeal of expedition. Scott and Amundson, Armstrong and Aldrin, Hillary and Tenzing, Livingstone in Africa, and Darwin on the *Beagle* — these *outdoorsmen* of high adventure I have followed, from dog-eared page to dog-eared page, and I have marveled at what they achieved. I have cast tears over the pitiful last paragraphs of Robert Falcon Scott, written as he lay in a snow-bound tent literally freezing to death; I have marveled at the tenacity and fortitude of Ernest Shackleton and his brave crew as they sailed, in a make-shift open boat, across the Antarctic sea to the island of South Georgia. I have mentally followed, step by weary step, climber Chris Bonnington to the summits of Mount Everest and Annapurna, and I have cheered on Thor Heyerdahl through my imagined variant of the *Kon-Tiki* expedition. I do not need to be close to death in order to feel alive. But in some strange voyeuristic way I can thrill on the adversity suffered and overcome by others — tall tales by tall people.

Is this a character flaw that I am admitting to? Quite possibly — I am not sure — I daren't explore. For all this, however, the *indoorsman* that I am has found excitement and marvel in the vicarious writings of adventure journals and stories, and I have most certainly felt the thrill and breathless moments of mental adventure and discovery. Journeys of thought, journeys of imagination — they require just as much struggle and resolve as that needed to climb Mount Everest, or to explore the Amazon rainforest or to set sail across the Pacific in a boat made of flimsy reeds and rope. The foot-slogging explorer certainly encounters more physical danger, and requires greater physical stamina, but the satisfaction and relief at journey's end is just as real for the former as it is for the latter. I do not need physical adventure, but I do appear to require and indeed, this is imperative, I do need to know, that someone has set out to conquer their fears and achieved those victories which I cannot.

Thought adventure — this is my domain of preference. I can brave its stormy seas and moments of doubt; I can navigate its boundaries and I can thrill at the discoveries made. Who needs the hardships presented by the real world, when the beauty of Plato's ideal realm is available for investigation? There within the world of perfect forms is my *indoorsman's* adventure playground. Thoughts, ideas, connections, numbers, and equations — neural adversity: this is a hardship I can usually deal with. Heinrich Hertz, who discovered radio waves in 1886, introduced the term *innere Scheinbilder* for the ability of the human mind to develop virtual mental images. Indeed, Hertz was one of the first scientists to develop a whole philosophical approach around the concept that mathematical systems (or sets of interacting equations) can stand for a representation of physical reality. In more recent times some researchers have taken the even more extreme stance that physical reality is no more, and also no less, than a mathematical structure. Physicist Paul Dirac, in a rare moment of personal disclosure, once commented that his research was not guided by mental models, but rather by mathematical beauty. This, of course, is a somewhat glib statement, presented no doubt *post hoc ergo propter hoc*. Beauty may be in the eye of the beholder, but beauty has no objective meaning and appearances can also be deceptive. In contrast to Dirac, astrophysicist Arthur Eddington, who graduated senior wrangler in the Mathematical Tripos for 1904, noted in his 1939 text, *The Philosophy of Physical Science* (Cambridge University Press), that "If I sometimes employ pure mathematics, it is only as a drudge; my devotion is fixed on the physical thought which lies behind the mathematics. Mathematics is a useful vehicle for expression and manipulation; but the heart of the theory is elsewhere." We see two very different attitudes being expressed towards mathematics, therefore, by two of the greatest scientific intellects of the 20th century. Physicists Stephen Weinberg is well known for his suggestion that anything that can be imagined must be realizable and that its physical counterpart must exist somewhere and some when in reality. A similar idea to this was introduced by T. H. White in his book concerning the young King Arthur, *The Sword in the Stone* (published 1938), in that at the entrance to an imagined ant colony was the sign, "everything not forbidden is compulsory." Weinberg's reasoning is ultimately sterile, however, since it invariably leads us to believe that the universe must be full of any number of weird and wonderful structures. Yes, the universe is full of weird and wonderful structures, but that is not because any specific human mind thought of them, and to echo the more sage words of evolutionary biologist John Haldane, "the universe is not only queerer than we suppose, but queerer than we can suppose." Even Haldane's statement is somewhat too bleak, however, since it implies that there are *things* going on in the universe that will never be accessible to scientific study and/or the human imagination. For all this, the thought experiment can take the human mind to places where the human body can never go. But once again we must be careful, and not everyone will be satisfied with the answer to a particular thought experiment that has been proposed. Austrian physicist Ernst Mach coined the term *gedankenexperiment* to specifically denote the imagined outcome of performing an actual experiment, and yet he strongly denied the existence of atoms since they could not be seen directly. The thought experiment, for us, is entirely built around the motivation of developing an understanding of the consequences of the principles relating to some very specific question. They are predicated on well-structured and well-known principles, and they are most definitely not flights of fancy — the laws of physics are not transmutable or specifically breakable in a thought experiment — rather they deal with the constraints of physics in an environment into which mind, but not the experimenter, can go — like that of an Earth-crossing tunnel.

Before one can imagine tunneling through the Earth, however, some clear knowledge of what the Earth physically is (its actual shape and extent) must first be put in place. Is the Earth a flat plate, is it a cube, is it an icosahedron, or what? In the mind's eye, the possibilities for the shape of the Earth are limitless, and yet, as we all know, the Earth is a sphere. The question now, of course, is how do we know the Earth is a sphere? Could it be, in the modern era of fake news and media lies, that the Earth is really banana-shaped, and that it is a conspiracy of world governments and scientists that have kept the greater public from knowing the plain truth? Well, yes this could conceivable be the situation, but with just a little rational thought about the question one can easily dismiss the conspiracy theory. At the very least, who would care if the Earth was actually banana-shaped, and why would governments from across the world, which can't even agree on essential matters such as basic human rights and civil decency, keep such secret knowledge... well, secret? The claim of a conspiracy concerning the spherical shape of the Earth is ludicrous; any yet many people hold and actively advocate such perverted views that the Earth is a flat disk or that we live on the inside of a great spherical shell. I will not waste time on picking through the problems of the flat- and hollow-Earth believers here — life is too short: *dare pondus indonea fumo* — but see Chapter 21. At this time, however, I will outline an experiment that anyone can reproduce to show that the Earth isn't a flat disk, that it isn't a rounded cylinder (an erstwhile perfect banana?) and that, in fact, it is indeed a sphere — or at least something very close to a sphere. The proof will be predicated on the present-day, all be they highly cynical, values of business and commerce — now here, surely, is a force that even conspiracy theorist can believe in [20].

As we all know a business does absolutely nothing for free, and the prime directive of any business management is to minimize production costs against profits. So, one can ask, in what manner do airlines fly their paying passengers from some specific point A to another specific point B? The answer to this question is simple, and it is the shortest possible path — such a flight path will minimize the amount of fuel required to transport the passengers, and it will also minimize the travel time, which means that the airplane can be put back into service, with a new set of paying passengers, as soon as the ground crew has refueled and revictualed the craft. With these criteria in mind, I carried out the following experiment on June 22nd, 2017, while seated aboard an Air Canada Boeing 777 aircraft flying from Calgary International Airport to London Heathrow. I was seated in the middle isle and conveniently placed right in front of my face was a display monitor. Primarily a device for advertising and in-flight entertainment, the display screen can also be made to show the aircraft's flight data, and over the eight-hour trip I kept track of the recorded flight time (that is the accumulated time in flight since take off) and the aircraft ground speed. Figure 8.1 shows a graph of the data that I jotted down, revealing the aircraft ground speed versus time of flight details. As one would suspect the aircraft reaches its characteristic cruising speed, about 950 km/hr, very rapidly. It is only within the first and last fifteen minutes that the aircraft ground speed is significantly below the average speed of about 821 km/hr. The impor-tant point about Figure 8.1, however, is that the area under the curve drawn through the data points corresponds to the total distance trave-led by the aircraft in its voyage from Calgary to London Heathrow — and this according to my number counts is 7,051 km. The question now is how does this deduced distance of travel compare with the various

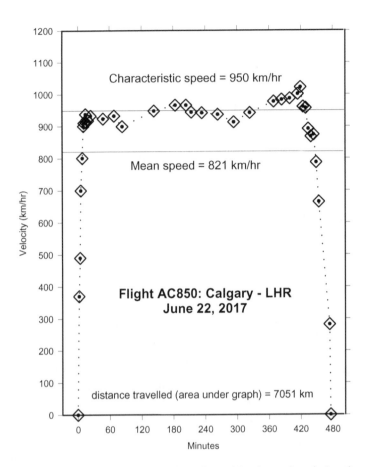

Figure 8.1. Speed versus time of flight data gleaned by the author during the AC850 flight on June 22nd, 2017, between Calgary and London Heathrow.

geometrical distances that can be constructed between Calgary and London Heathrow?

In terms of geographical longitude, Calgary and London Heathrow are $\Delta\lambda = 114$ degrees apart; additionally, it turns out, usefully for this experiment, that Calgary and London Heathrow have the same latitude — each being $l = 51$ degrees north of the equator. With this information we can accordingly calculate three specific distances: (i) the straight-line distance L between Calgary and London Heathrow, (ii) the cylindrical, constant-latitude distance C between Calgary and London Heathrow, and (iii) the

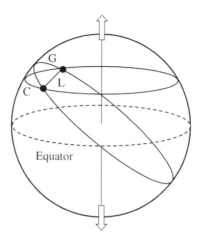

Figure 8.2. The three geometrical distances that can be drawn between two locations (solid dots) in space. *L* is the linear distance (according to a flat Earth), *C* is the cylindrical distance and *G* is the great circle distance.

great circle distance *G* between Calgary and London Heathrow. The last measure is perhaps less well known in everyday life, but the great circle path corresponds to the circle, centered on the Earth's core, which passes through the two locations of interest, Calgary and London Heathrow in our case. Figure 8.2 indicates the various distance measures being considered.

Taking the Earth's radius to be $R = 6,371$ km (a number that we shall justify, again through direct experimental measurement, later) we have that the following: $L = 2R\cos(l)\sin(\Delta\lambda/2) = 6,725$ km; $C = 2\pi(\Delta\lambda/360)$ $R\cos(l) = 7,977$ km; and $G = 7,023$ km. The great circle distances *G* must be calculated through the application of spherical trigonometry formula and the details, while readily available, are skipped over at this time. Comparing these distances with that flown by flight AC850, we find that the aircraft flew a distance of 365 km more than the direct linear path, a distance 926 km less than the path of constant latitude (the cylindrical path), and a distance just 28 km longer than the great circle path between Calgary and London Heathrow (we did have to go into a holding loop before landing at Heathrow). Applying now our commercial parsimony rule of traveling the shortest distance possible, it is clear that the aircraft has followed the great circle path, which is indeed, the shortest distance between any two points located on the surface of a sphere — accordingly, aircraft flight times are fully consistent with the Earth being a sphere, but not with the Earth being either flat or cylindrical. If the Earth were flat then airlines (and passengers) are certainly paying for more fuel than they need to, and the flight time between Calgary and London Heathrow should be much shorter, for the given aircraft speed, than recorded. The Earth is a sphere (or spheroid) — no question.

Chapter 9

Tik-Tok's Tumble

Logic will get you from A to B. Imagination will take you anywhere.

A. Einstein

Aesop's Fables date back, at least in oral tradition, some two-and-a-half thousand years. They are old and wise tales, usually having some form of moral twist, which have, over the centuries, been added to and adapted according to varying cultural tastes. One of Aesop's more recently constructed fables concerns an astronomer, and it runs something like the following:

An aged astronomer was out walking across a field one night, and, with his attention fixed upon the sky above, accidentally fell into an open well. After many hours of calling for help, a passer-by chanced to hear his cries, and upon learning what had happened decried, "why you silly old fool, in striving to learn what is happening in the heavens, you have failed to see what is happening here on Earth."

The supposed moral of this fable is that one should not look so high as to miss seeing the things that are right in front of you. I am not sure I fully agree with the literary mechanics of this particular fable — for indeed, an astronomer is supposed to be looking up at the sky when actively studying the heavens. Perhaps the moral of the fable is more attuned to the idea that astronomers should not *walk* in a field where there might be deep wells while out observing the sky on a dark

night — but this is a less helpful result to the everyday reader.[1]
Nonetheless, for all that one might criticize the supposed moral behind
Aesop's astronomer's fable, the premise of accidentally falling down a
well, or some deep cavern, has, more often than not, been the starting
point for the majority of stories concerning the exploration of the Earth's
interior. Indeed, few such exploration stories are predicated upon a
deliberate plan of action to investigate the Earth's interior. Rather, it is
usually an accident, or some form of mechanical failure, or some act of
betrayal, which determines the eventual path of the subterranean
explorer. Indeed, there is a strange, almost eerie, lack of literature
regarding underground exploration. Many a literary hero has been type
cast as a voyager upon the open seas, or as an explorer of outer space,
or as a navigator of alien worlds, but the idea of having a hero set out to
specifically explore the inner Earth is something that relatively few
authors have attempted to do. Why is this? The answer is no doubt complex, but may be partially due to the fact that the oceans are big and
expansive, dotted with exotic islands and strange wild animals and sea
creatures; likewise space is expansive and full of light and populated by
fully imaginable worlds of one form or another. The inner-Earth, in contrast, is sunless, dark-filled, restrictive and overpowering. Long held as
the place of the dead, the underworld is an oppressive mass of rock and
boiling magma and isolation. Is it not strange, in fact, that one of the
great mantras of our time is that space is the new frontier? Oddly,
indeed, we know more about the Sun, Moon and the workings of the
stars than we do about the abyssal oceans and Earth's deep interior.
History seems to be telling us that few authors have dared to think of
subterranean exploration as being anything other than an act of calumny
and/or a nightmare journey haunted by dinosaurs and demons. The
inner-Earth, it would seem, is a place where no human being should will-
ingly go either physically or mentally.

The crossing of the bridge of Khazad-dûm (the *black chasm*) represents a pivotal moment in J. R. R. Tolkien's masterpiece *The Lord of the*

---

[1] There are fabled reports of hapless astronomers, upon returning from a night's observing at the Royal Greenwich Observatory, when it was located at Herstmonceux Castle in
East Sussex, of falling into the decorative ponds that were inconveniently and unneces-
sarily set, as architectural features, between the various telescope domes.

Rings (first published in three volumes between 1954 and 1955). It provides a dramatic backdrop to the fight between good, in the form of Gandalf the Grey, and ancient evil, in the form of a fire-breathing Balrog. Furthermore, with the eventual killing of the Balrog, the bridge-crossing episode is pivotal in setting in place the resurrection/transformation of Gandalf, through the action of the Valar, into Gandalf the White. The passage across the bridge additionally witnesses the escape of the ring bearer and companions from the mines of Moria, thereafter finding safety in Lothlórien. It is the prelude to the sundering of the fellowship. The span is described as being "a slender bridge of stone, without kerb or rail, that spanned a fifty-foot wide chasm of indeterminate depth." Indeed, Gimli the Dwarf recounts that "deep is the abyss that is spanned by Durin's Bridge, and none has measured it." While described as a defensive feature, over which any would-be army attacking Moria must pass, the bridge is also a metaphor for the Christian life journey — stray from the righteous path by just a small amount, for indeed the path is narrow, and you will fall; the eternal soul becoming lost in the equally infinite depths of shadow. The bridge is the narrow path of safety, and Gandalf's fall implies eternal loss — the reader is literally left in shock and despair at the (apparent) death of one of the storie's major and most-wise characters. Tolkien, of course, had other plans for his hero. "Long time I fell," recounted Gandalf later, "then we plunged into the deep water and all was dark. Cold it was as the tide of death: almost it froze my heart." Gandalf's long fall is ended by encountering a deep subterranean ocean/lake, in similar ending, in fact, to many of the other stories about deep-Earth travel that we shall encounter below. Gandalf and the Balrog do not fall to the center of the Earth, but continue their battle through numerous dark caverns and interconnected tunnels, reminiscent in description to the Swiss-cheese inner-Earth picture propounded by Athanasius Kirchner in the 17th century. Eventually, however, Gandalf finds his way to the "Endless Stair [...] ascending in unbroken spiral in many thousand steps." Gandalf's fall is steeped in religious symbolism; the fall into darkness is long and fraught with danger and despair, and it carries the body to the oppressive depths of the deep-Earth and a long entrapment in Hell. There is always hope, however, that evil can be overcome. The journey from the dark depths, back to the light and Earth's surface, is interminable and paved by a near-endless, winding stair — reminiscent, one suspects, of climbing out of Flamsteed's well.

Indeed, Gandalf's fall through the abyss is steeped in elements of theology, Dante's Hell, and Jules Vern's 1864 science-fiction classic, *Journey to the Center of the Earth*. Tolkien would certainly have been well aware of and indeed well versed in Dante's nine levels of Hell. Forming part of Alegea Dante's 14th century epic poem the *Divine Comedy*, the *inferno* recounts the journey through the nine circles of Hell by Dante and his guide the ancient Roman poet Virgil. "Abandon all hope, ye who enter here" is the greeting at the first gate of Hell, although Hell proper is not encountered until Charon arrives to ferry our erstwhile explorers across the subterranean river Acheron. The various circles of Hell are arranged so as to accommodate the nine deadly sins, and at the center, at the very core of the Earth itself is *Judecca*, a frozen wasteland of ice in which the treacherous are entombed. Here, frozen waist-deep in ice, Dante and Virgil find the railing and ever-tormented Lucifer. In this depiction Hell has indeed frozen over, and the core of the Earth is taken to be a frigid emporium. Dante and Virgil continue the journey from the center back to the Earth's antipodean surface, taking a total of 2 days to complete their journey cross its entire diameter. The climb out from the core to the southern hemisphere surface is, in fact, remarkably quick, taking our intrepid explorers a mere one-and-a-half hours.

Of all the texts concerning subterranean travel, Jules Verne's *Journey to the Center of the Earth* is by far the best known — it has been continuously in print for over 150 years and has been made into numerous movies. It is also incorrectly titled, since the hapless explorers do not actually travel to the center of the Earth. Verne wrote his story at a time when the science of geology was undergoing rapid change and development. Starting with Charles Lyell's *Principles of Geology*, published in 1830, and Charles Darwin's *Origin of Species*, published in 1859, by the time Verne began his story it was clear that the Earth, rather than being the biblical 6,000 years old, as calculated by Bishop Ussher,[2] must be at least many millions of years old. It was this ancient Earth that Verne wanted to

---

[2] James Ussher (1581–1656), Archbishop of Armagh in Ireland, calculated from a literal reading of the *Old Testament* that the Earth was created at about 6 pm on October 22nd, 4004 B.C.

describe, and so it was that Professor Otto Lidenbrock, his nephew Axel, and Icelandic guide Hans, were created to follow in the footsteps of the first deep-Earth explorer Arne Saknussemm: "Descend, bold traveler, into the crater of the jökull of Snaefell [...] and you will attain the center of the Earth. I did it." As with all of Verne's books, the plotline to the *Journey* is imaginative, but the end literary result is rather weak, contrived and unsatisfying. For all this, the various twists and turns of the adventurers eventually find them on the shores of a subterranean ocean situated some 100 miles below ground and beneath the Grampian Mountains of Scotland. Following one mishap after another, the adventurers eventually chance to fall through a whirlpool of water that takes them, after many hours, at a speed of "not less than a hundred miles an hour," to another interior ocean and cavern. Having thus trapped his fictional characters to starve underground, Verne writes a highly contrived ending, with his heroes making an escape on an up-rising column of magma — the adventurers eventually gaining the Earth's surface upon being vented from Mount Stromboli to the north of Sicily. Verne's *Journey* is a contradictory read. On the one hand Verne employs detailed scientific dialog and reasoning, but on the other, mixes fantastic beasts and the tall tales of ancient explorers into his narrative. This latter material, of course, is intended to make for a ripping good yarn, and after all it was not Verne's intention to write a book on geology. What is perhaps most important about Verne's *Journey*, however, is that it opens up, albeit in a fantastic sense, the idea that the deep-Earth might be deep-searched through scientific reasoning and methodology. When, in 1864, Verne wrote of the quest to follow the ancient trail laid down by Saknussemm, there was no actual consensus as to whether the Earth's interior was hot or cold, or whether there really was a subterranean ocean, and it was to be at least forty years before the very first seismographs were to be deployed in order to listen in on the Earth's quaking mantle.

A story in similar vein to Verne's *Journey to the Center of the Earth* is that of *A Strange Manuscript Found in a Copper Cylinder* by James De Mille.[3] This latter book, first published in serial form in 1888 (although probably written in the 1860s), is a much superior read to Verne's, and

---

[3] Closer parallels to De Mille's text are Arthur Conan Doyle's *The Lost World* (published in 1912) and *The Land that Time Forgot* by Edgar Rice Burrows (published in 1924).

the narrative is all the more imaginative and griping. Once again it is the tale of a journey, in this case starting in the Antarctic Ocean, and the passage through a vast subterranean tunnel, only to end in a sub-tropical south polar sea. The fictional writer of the manuscript, Adam More, reminds the reader early on in his narrative, "as a boy I had read wild works of fiction about lands in the interior of the Earth, with a Sun at the center, which gave the light of a perpetual day. These I knew were only the creations of fiction; yet, after all, it seemed possible the Earth might contain vast hollow spaces in its interior." De Mille's text is mostly concerned with the dystopian world of the Kosekin, but it is predicated upon the notion that the inner and indeed, the very ends of the Earth are (or were then) *terra incognita*. In Verne-like fashion, De Mille includes dialog that explores the details behind the measurement of the Earth's oblateness, refereeing specifically to the work of James Ivory, and he also re-introduces to the world dinosaurs, ichthyosaurs, pterodactyls and dodos. De Mille essentially exploits in his text the uncertain characteristics of the actual Earth, at that time, as a foil against which to set his imagined dystopian world and its fantastic creatures. Edgar Rice Burrows continued the theme of populating an imagined inner-world with ancient creatures in his *At the Earth's Core* (first serialized in 1914, and published in book form in 1922). Burrows enhances his narrative, however, by introducing the now common science-fiction trope of technology failure. Indeed, the inner surface of the Earth's shell, Pellucidar, is revealed through the loss of steering control aboard an experimental mining machine called the Iron Mole. Once set in motion the machine heads irreversibly downward and after 500 miles of descent breaks through to the surface of Pellucidar. This inner-shell world is ruled over by intelligent flying reptiles, the Mahars, who keep in thrall a population of stone-age-like humans. The story is not a great work of fiction, but it develops the literary theme of populating possible worlds with ancient beings and alternate histories compared to that played out on Earth's surface. Once again, as with Verne's *Journey*, we also find a title that is at odds with the story line in that the core of the Earth is, in fact, not revealed and/or encountered. Indeed, Pellucidar, spread over the inner surface of the Earth's crust, is kept in permanent daylight by a central Sun. There is also a Moon that occupies a fixed position with respect to the central Sun which casts a

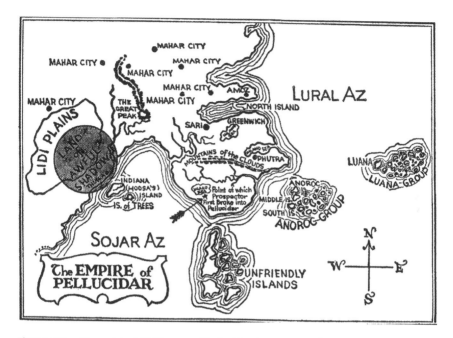

Figure 9.1. The original 1915 map of Pellucidar (1962 reprint). The arrow (center left) shows where The Prospector (Iron Mole) broke through into Pellucidar. The circle at center left indicates "the Land of Awful Shadow."

zone of permanent darkness on Pellucidar called the Land of Awful Shadow (Figure 9.1).

Burrows was far from the first writer to develop a gravity-defying inner-Earth dynamical system. Indeed, Dano-Norwegian philosopher and essayist Ludvig Holberg wrote in his endearing novel *Niels Klim's Underground Travels* (first published in Latin in 1741) of a miniature planetary system within the interior of a hollow-Earth. As with virtually all texts concerning the discovery of a hollow-Earth, the narrative begins with a disaster. While exploring a deep cave in Bergen, the rope supporting the hapless explorer breaks, "and I tumbled with strange quickness down the abyss," explains Klim (the story's narrator). We then learn that "enveloped by thick darkness, I had been falling about a quarter of an hour, when I observed a faint light and soon a clear and bright-shinning heaven. I thought, in my agitation, that some counter current of air had blown me back to Earth." It was not so, however, and our narrator soon realizes that he has fallen into a subterranean firmament, "the Earth is hollow, and within its shell there is another, lesser world, with corresponding suns, planets, [and] stars". Mysteriously, Klim's rate of descent is slowed, and upon penetrating the atmosphere of one of the inner-Earth planets, he goes into orbit around it, and there he remains for three days. Eventually, after being attacked by a giant eagle, Klim is brought to ground on the planet Nazar.

Needless to say, the inner worlds composed of suns, stars, planets, and geometrically fixed Moon, as envisioned by Holberg and Burrows, are physically impossible structures — the dynamical stability (let alone the formation) of such configurations being denied through Newton's shell theory.[4] In several ways, however, Burrows anticipates the idea of the Dyson Sphere,[5] in which it is envisioned that a technically advanced society might build a complex of orbiting satellite-worlds around its parent star in order to tap more of the radiant energy — such a structure was "encountered" in the *Star Trek: The Next Generation* episode *Relics*, first aired in 1992. Once again, however, a coherent, rigid shell enveloping a star would not be dynamically stable — numerous island satellites, however, provided there is no appreciable gravitational interaction between their numbers, can remain on stable orbits around a central star.

Several film adaptations of Burrows' *At the Earth's Core* have been made, but none have acquired critical acclaim. Indeed, the most recent version, *The Core* (2003: Directed by Jon Amiel and released by Paramount Pictures) was voted in the *MIT Technical Review* for June 2011, the second worst science-fiction film ever made. While the distain of the *Review* was directed at the clumsy portrayal of the film's technical

---

[4] The dynamical stability here is not related to a planet orbiting around a supposed inner sun, nor a moon in orbit around a supposed inner planet. Rather it relates to the fact that if the central star was to shift, even ever so slightly, from the exact center of the Earth-shell, then since there is no restoring force (as indicated by Newton's shell theory) it would simply keep moving until it crashed into the shell's inner rim — presumably with disastrous consequences for the inner and outer-Earth inhabitants.

[5] Introduced by Freeman Dyson in 1960, such mega engineering projects have now become a target for SETI. Being able to complete such a major engineering program is taken to be the sign of a so-called Kardashev type II civilization. Humanity presently qualifies as Kardashev type I civilization.

and scientific content, the premise of the film was actually concerned with the core of the Earth. Indeed, Earth's inner core had stopped rotating (a physical impossibility) and with the concomitant breakdown of the Earth's inner-dynamo its protective magnetic field was in danger of collapse. To start the core rotating again, a human-controlled, Iron Mole-like capsule is developed to deliver a nuclear bomb to the inner regions of the Earth — again, to the assembled groans of the science reviewers, the craft was fashioned out of a very rare, high-pressure, high-temperature resistant material called "unobtainium." For all this bickering, however, *The Core* is at least premised on a realistic model of the Earth's interior, although no specific passage through the Earth is portrayed.

In stories subsequent to *At the Earth's Core* Burrows allowed access to Pellucidar through large openings located at Earth's polar caps — in contrast to the sub-tropical zones envisaged to reside there by De Mille. The idea of access to the Earth's inner regions through polar portals has a long and tortuous history. Perhaps the best (of a bad lot — see Chapter 21) description for such a hollow-Earth model is that developed by Marshal B. Gardiner and described in his 1913 text *A Journey to Earth's Interior*. Indeed, Gardiner secured a U.S. Patent, number 1,096,102, for his Geophysical Apparatus in May 1914. This device constitutes a standard terrestrial globe on the outside, but can be opened out to reveal details of the Earth's inner shell and a centrally located star (represented by a glowing lightbulb) (Figure 9.2). In Chapter 21 we shall encounter the persona of John Cleves Symmes, amongst others, who made the hollow-Earth, with polar openings, idea both popular and controversial; Gardiner, however, tried to set himself apart from Symmes and similar such practitioners, and he attempted to develop a scientific theory based upon the *facts*[6] to explain why the Earth, and indeed the Moon and any other planet, should form as hollow structures containing an inner star — his approach was at least different to the norm in which knowledge was simply handed down by divine decree.

---

[6] The *facts* employed by Gardiner are, of course, selective, inconclusive and poorly interpreted. His thesis is essentially premised upon a mass of confused arguments, wishful thinking and illogical reasoning.

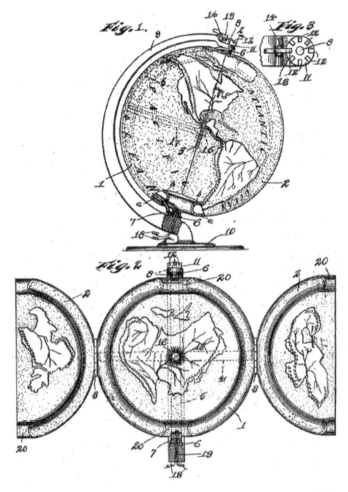

Figure 9.2. U.S. Patent, number 1,096,102, showing the design of Marshal Gardiner's hollow-Earth globe. The application states: "Upon the outer surface of the globe [Fig. 1] are the usual geographical indications or maps illustrating or indicating the continents of the world. Upon the inner surface of the globe are also arranged geographical indications [Fig. 2] illustrating continents which according to the theory of the inventor exist upon the inner surface of the globe."

The first story to use the idea of passing through the Earth, from one pole to the other, as a plot device appears to be the 1721 publication, in France, of *Relation d'un voyage du pôle arctique au pôle antarctique par le centre du monde* — literally, *A Journey from the Artic Pole to the*

Antarctic Pole via the Center of the Earth. It is not actually known who composed the text, but it is a rather strange tale with the adventurers seemingly having the habit of killing and then eating all of the new animals and birds that they encounter. The title though is somewhat misleading in that no detailed description of the passage through the Earth is given, and the story is essentially based on the portal-passage genre where explorers are carried from the known to the unknown via some unexpected gateway. The gateway in this case is a tunnel that takes the adventurers from the then semi-explored Arctic to the then entirely unexplored Antarctic. The journey is precipitated by the encounter of a "frightful whirlpool at the North Pole [...] in the middle of which there has to be a frightful bottomless gulf, into which all the waters of the seas precipitated, communicating via the center of the world with the seas at the Antarctic Pole."[7] The explorer's ship having being caught in the spiraling gyre is boarded up by the crew, and they wait out their ordeal inside: "for whatever heaven had ordained." As for the journey through the Earth, it is covered in two sentences: "we felt ourselves sinking into that profound abyss with an inconceivable rapidity. The horrible whistling and buzzing that we heard around us incessantly, importing terror into our souls." With this the crew falls into a deep faint only to awake alive and well, with their faithful ship gently bobbing in the calm seas off the southern Pole. The rest of the story is then a fantasy adventure tale, with the explorers encountering one strange beast or artifact after another. Even the great Baron Munchausen, renowned for his loquaciousness, gave small shift to his passage through the Earth.

Invented under the sharp-witted pen of Rudolph Raspe in 1785, the Baron, having brought the jealous wrath of Vulcan upon his head, is thrown down the throat of Mount Etna, "I found myself descending with an increasing rapidity, till the horror of my mind deprived me of all reflection." Upon being suddenly plunged into water, however, the Baron reasoned, "I had passed from Mount Etna through the center of the Earth to

---

[7] It is worth noting that in 1721 the idea of a large continent residing at the South Pole was no more than speculative myth. The first accredited sightings of the Antarctic continent were only made in the 1820s, and even in 1869 Edward Everett Hale in his short story on early space exploration, *The Brick Moon*, can have his observers, without fear of contradiction, comment that, "Your North Pole is an open ocean... Your South Pole is an Island bigger than New Holland. Your Antarctic Continent is a great cluster of islands."

the South Seas. This, gentlemen, was a much shorter cut than going round the world, and which no man has accomplished, or ever attempted, but myself: however, the next time I perform it I will be much more particular in my observations." Raspe was an accomplished geologist and a mining expert of his time and it is perhaps surprising that he was seemingly unprepared to even speculate upon what the good Baron might have witnessed during his plunge through the Earth. There was no such laconic response, however, when some 130 years after his fall through the Earth the Baron fell through the Moon. *The Scientific Adventures of Baron Munchausen* was written in serialized form by Hugo Gernsback[8] starting in 1915. The *Scientific Adventures* were written in the form of radio installments and updates broadcast by the Baron from the Moon (and later from Mars). That the Baron survived passage through the Moon is revealed as consequence of a near disaster. "A meteor [*sic*[9]] crashed down on my aerial 50 feet from where I was sitting [...] and I was blown [...] right down into the mouth of a giant crater, by the colossal resulting blast of concussion," the Baron informs his radio listener. That the crater had no bottom, the Baron continued, was evident because he could see several stars shining through from the night side of the Moon.[10] The Baron realized that he was then falling through the Moon, and with great calmness of mind records that "I knew the diameter of the Moon to be 2,164 miles. A quick mental calculation proved that it would take my falling body about 24 minutes to reach the center of the Moon." The text then goes on to reveal that unless he could engineer some escape the Baron would be doomed to oscillate back and forth across the Moon's interior — in an exact parallel to Terricola's prize problem described earlier. Clearly, Hugo, it would appear, was aware of the mathematical solution to the Earth-tunneling problem, although he miscalculates the travel time. We saw earlier the travel time T across the interior

---

[8] This is the same Hugo Gernsback after whom the prestigious annual Hugo Award for best science-fiction writing is named.

[9] Technically the term meteor corresponds to the light phenomenon associated with the destruction of a meteoroid in the Earth's atmosphere.

[10] This, in the light of our earlier discussion of seeing stars up chimneys, is entirely impossible.

of a uniform density body is given by the expression $T = \pi(R/g)^{\frac{1}{2}}$, where R is the radius and g is the surface gravity (see also the Appendix). Taking the Moon's gravity to be 1.6 m/s^2, then the time to reach the center would be 27 minutes (that is, half of $T = 54.3$ minutes).

An interesting twist is added to the Baron's flight across the Moon's interior, and indeed a twist that negates the parallel with a passage through an Earth tunnel. The text explains that after his miraculous escape from his predicament, the Baron determined that the Moon was not solid, but an immense hollow shell, "with a solid crust about 500 miles thick." The text explains that the Moon's interior had been hollowed out by the centrifugal forces that were acting over time before it solidified, "some millions of years ago." Here Gernsback is seemingly mixing incorrect physical reasoning, with ideas similar to (but in reverse effect) those introduced by Marshal B. Gardiner in 1913. Gernsback's notion also appears to build upon the fission origin for the Moon as introduced by George Darwin in the early 1900s. There are also echoes of Edward Hale's *The Brick Moon*, published in 1869, in Hugo's thinking. This latter text is often described as being the first science-fiction story in which artificial satellite technology and space exploration is implicitly developed. Indeed, Hale's idea was to place such moons in what would now be called geostationary orbits so as to aid in the determination of longitude at sea. By arguing for a hollow-Moon, however, the Baron (that is Hugo) voids an exact parallel with the Earth-tunneling problem — indeed, mathematically it is a more interesting problem, as revealed in the Appendix for those readers who wish to see the details.

At about the same time that Hale was placing brick moons in Earth orbit, English mathematician Charles Dodgson (better known under his penname of Lewis Carroll) envisioned a small girl falling down a rabbit hole. Indeed, while not explicitly developed, Dodgson's *Alice's Adventures in Wonderland* (first published in 1865) is set in a subterranean world: "Down, down, down. Would the fall never come to an end? 'I wonder how many miles I've fallen by this time?' she said aloud. 'I must be getting somewhere near the center of the earth?'" We do not find out from Dodgson just how deep the rabbit hole goes, but Alice eventually comes to a soft landing in a fully illuminated and peopled, albeit odd-peopled, world. As a mathematician we might well expect Dodgson to have been aware of the Earth-tunneling problem (as set by Terricola) but he makes no play of its possibilities in Alice's adventures. In the first of his later stories featuring Sylvia and Bruno, however, Dodgson introduces the reader to Professor Mein Herr, and this august gentleman is not only from another planet, but he knows all about gravity trains (see Chapter 17).

Writing just one year before Gernsback's first installment of the Baron's *Scientific Adventures*, Lyman Baum published in 1914 his charming tale of *Tik-Tok of Oz*, featuring Betsy Bobin, Hank the mule and, of course, Tik-Tok him (or more correctly, it) self. Indeed, Tik-Tok is a fully dehumanized version of the Tin Man of Oz; built from the ground up, as it were, by Smith and Tinker as a mechanical man/robot, rather than being re-created limb-by-limb, as in the case of the prosthetic-gaining Nick Chopper. Tik-Tok is not alive but driven by clockwork — indeed three pieces of clockwork, one each for its thoughts, its actions and its speech. A copper plate on the automaton's body reveals that it "Thinks, Speaks, Acts, and Does Everything but live" — another engraved plate indicates, "This machine is guaranteed to work perfectly for a thousand years" (now we truly know that we are in the land of make-believe). For all this, provided that someone is prepared to turn its windings, Tik-Tok is a loyal and powerful companion. In *Tik-Tok of Oz*, our erstwhile robot, in spite of the story title, hardly features at all, but nonetheless accompanies Dorothy and the Shaggy Man on their quest to release the latter hobo's brother who has been imprisoned by the Nome King. The story is characteristic of the Oz family of writings, and the only chapter of interest to us here is Chapter 10: *A Terrible Tumble Through a Tube*. As appears to be the literary standard, Baum's characters fall unexpectedly into the Earth-tube: "she [Betsy] could see nothing at all, nor could she hear anything except the rush of air past her ears as they plunged downward along the tube." We are told that the company fell "for more than an hour" and that the "trip was a long one, because the cavity led straight through the Earth." The tunnel was, we are additionally informed, built by Hiergargo — a magician — who was a great traveler, and who reasoned that going through the Earth was quicker than going around its surface. From the story details provided by Baum, it is clear that he was unaware of (or at least unencumbered by) the physical workings of the Earth-tunneling problem. Indeed, its effect upon the motion of the magician, Tik-Tok,

Betsy and Hank is more aligned with that of cartoon gravity[11] than that of the real world or within our idealized problem. This situation is in complete contrast to the serialized short story *Through the Earth*, first published in the *St. Nicholas Magazine* in 1898, by Clement Fezandie.

Hugo Gernsback described Fezandie as a "titan of science fiction" and *Through the Earth* was one of his earliest of contributions to that genre — and it's a belter. The text is brimming over with science-fiction speculation, not only about how one might build a tunnel through the Earth, but about the machinery required to do the job, and the material out of which the tube should be made (a newly invented material called carbonite — we are told), and the fact that if the transit tunnel is going to work at all then its interior must be maintained as a sealed vacuum. Not only this, Fezandie drives his tunnel through what is a decidedly modern-sounding Earth, completed with a hot molten core — although the alternative of a solid interior is debated. The story is essentially a dialog between a naïve questioner, James, and an authoritative Dr. Joshua Giles. Indeed, it is Giles that provides all the scientific answers to the narrator's (as well as the engaged reader's) questions. Well, of course, a story wouldn't be a story if it just revealed the possibility of building a transit tunnel through the Earth, and disaster, of course, strikes during the maiden voyage of the passenger cabin. While no life is lost in the narrative, the tunnel is destroyed — in spite of its cooled carbonite shell. In a remarkably abrupt ending the story concludes: "As for the transportation company Dr. Giles had organized, I regret to say that it was dissolved, as the dangers from the central heat of the Earth were found to be too great to be risked with impunity."

The essential ideal behind the construction of the transit tunnel in Fezandie's *Through the Earth* was commerce. It would enable a rapid exchange of goods between the southern and northern hemispheres, and money was also to be made from the material and ores extracted in the construction process. These very same ideals, with a slight nod

---

[11] In the cartoon universe, gravity only operates some time after a character has walked over the edge of a cliff and chanced to realize their situation. Douglas Adams also exploited this idea in his *Hitchhiker's Guide to the Galaxy* series of books — the ability to fly is mediated by the (difficult to master) action of forgetting to hit the ground after deliberately falling over.

towards the scientific advancements that would follow, were outlined by Camille Flammarion in a truly wonderful article, aptly titled *A Hole Through the Earth*, published in *The Strand Magazine* for September 1909. Flammarion is largely remembered now for his popular writings that combined aspects and ideas from science, science fiction and spiritualism. His *Strand Magazine* article, however, is set very much in the middle realm between science and science fiction. Flammarion writes that he has had the idea of "sinking a shaft into the Earth for the express purpose of scientific exploration" for some time, and that this shaft should be cut to a depth "as far below the surface as the utmost resources of modern science would permit" (Figure 9.3). His motivation for writing the article, Flammarion explains, was the occurrence of recent earthquakes[12] and "the extremely contradictory opinions of geologists upon the interior state of the Earth." Indeed, at the beginning of the 20th century the geological world was still divided as to whether the Earth was more or less solid within its interior, or whether, in Flammarion's words, consisted of a "liquid and incandescent core contained within a thin shell or crust." To resolve this issue Flammarion proposes an entirely pragmatic approach — dig down, and dig deep. In this latter respect Flammarion is being entirely practical and envisions a vast army of workers and machines pushing ever deeper into the Earth's interior. The financial benefits of this project, Flammarion points out, will be in the form of extracted minerals (iron, copper, gold, and "elements hitherto unknown and unexpected"). The scientific gain, of course, would be in the study of the Earth's interior rocks and in the unraveling of the mystery behind the increase in temperature with depth. Is the Earth's interior heat primordial in origin, as advocated for by Lord Kelvin, or is the heat derived from the decay of radioactive elements, as advocated by Lord Rutherford? Not only would the construction of an Earth-penetrating shaft yield financial wealth and scientific knowledge, Flammarion also argues that it could solve national unemployment and stop war. Indeed, Flammarion suggests that the tunnel should be constructed entirely by soldiers, convicts and the otherwise unemployed. Flammarion doesn't actually envision a tunnel being cut through the entire Earth, but does, at the end of his

---

[12] The great San Francisco earthquake, for example, occurred in 1906.

Figure 9.3. A construction crew and heavy machines begin to cut a tunnel through the Earth. Image from the *Strand Magazine*, September 1909.

article, consider the question of "what would happen to a body falling into such a shaft." In a tunnel of finite depth, of course, the answer is that they would die horribly and literally splat apart as they hit the tunnel floor. Flammarion, however, chooses to consider the Earth-penetrating tunnel problem (Cymro's problem), and correctly indicates that the hapless body would return to the tunnel entrance into which they had fallen after a time of 84 minutes: "[they] would thus continue to describe a series of oscillations like a new kind of pendulum."[13]

The mechanical, hard-graft, super-machine approach of exploring the Earth's interior, as advocated by Flammarion, never came to pass,[14] and indeed, the necessity of physical tunneling to determine the Earth's internal structure was negated by new developments in seismology in the early 20th century.

With little doubt the most brazen literary application of the hollow-world idea is that by American author Edmond Hamilton — a prolific writer who in his later life wrote numerous Batman and Superman stories for DC Comics. In his early stories, however, Hamilton was concerned with the adventures of Captain Future (a.k.a. Curtis Newton, "man of tomorrow," who was born and raised on the Moon). Of specific interest here is the story of *Outlaw World* (first published in 1946). In this tale of space adventure and daring-do Captain Future battles against a band of space pirates and tracks them to their secret base set within the interior of planet Vulcan (Figure 9.4). This is a wonderful melding of literary science fiction with historical science fiction. For indeed, planet Vulcan was written into existence by the mathematical calculations of Urbain Le Verrier in 1859. In his detailed accounting of gravitational interactions, Le Verrier found that he could not explain the observed rate of perihelion advancement of Mercury. As a consequence of this theory — observation mismatch, Le Verrier, certain in the correctness of Newton's equations, invoked an inter-Mercurial planet to account for the missing variation — a miniscule 43 seconds of arc per century. This planet was given the name Vulcan (in honor of the Roman god of fire), and for many decades, stretching both sides of 1900, Vulcan was the target of many astronomical searches — all of which failed (as they must) to find the required additional planet — which is not to say that claims of detection weren't made. It was not until 1915, when Albert Einstein introduced his general theory

---

[13] One might say that this gives a whole new meaning to the term pendulum bob — first he was our eponymous uncle and now he is some kind of high-flying, come-again, gone-again laborer.

[14] We do note, however, that Flammarion's *Strand Magazine* article was reproduced in the October 2nd, 1909 issue of *The Advertiser* in Adelaide, Australia — so at least it can be argued that his article made the trip from one hemisphere to the other.

Figure 9.4. The hollow world of planet Vulcan. The interior is illuminated by sunlight that shines through a large opening (called The Pit) and the interior (which supports an atmosphere!) is blanketed by a lush carpet of jungle growth. The location marked "comet landing" indicates where Captain Future's spaceship, the *Comet*, first touched down in the interior.

of relativity, that the perihelion advancement of Mercury was fully explained, and just as Le Verrier's pen had written Vulcan into existence, so Einstein's pen, 56 years later, crossed it out. Hamilton, in his *Outlaw World*, however, uses both the mystery and mystic of planet Vulcan to advance his storyline; a storyline that would likely still have some resonance in the vaguely recorded memories of at least his older readers. The fact that Vulcan is taken as being hollow and habitable on the inside (and populated by giant, crab-eating "rock apes") adds to its strangeness, and it also acts as a literary counterfoil to the misguided world views expressed by such writers as John Cleves Symmes and Cyrus Teed (see Chapter 21).

Perhaps one of the first questions that an engineer might ask, apart from that concerning the sanity of the proposer, if approached on the issue of excavating a tunnel through the Earth, is "how long would such a tunnel be?". This, of course, is the diameter of the Earth and to answer how far this might actually be, one could begin by digging a small vertical tunnel, a veritable well in fact. Indeed, it was this very approach that Eratosthenes made use of in the 3rd century B.C. to derive the first physical measure of the size of the Earth — as we shall see in the next chapter.

Chapter 10

Eratosthenes's Well

When the Sun sets, shadows that showed at noon
But small, appear most long and terrible.

Oedipus, Nathaniel Lee (1679)

It is almost impossible, from a modern perspective, to envisage life in ancient Greece. The more-than-2000-year gap between then and now presents us with an insurmountable barrier. We catch only shadows and reflections of what it was like to be a citizen of that ancient world. Something approaching a human perspective and appreciation can be gleaned, however, from the snatches of texts and the translations of translations of texts that have survived to the modern era, but we are largely left with just snippets and rumor. We find and read about people who are only partially complete; we do not know them, we only know their works, or more often than not, rumors of their works that were produced by later commentators. For all of this lack of human contact, however, a remarkable data set of ancient ideas and speculations upon the cosmos has been pieced together by near uncountable scholars and translators. We do not know the individuals but history has given us their thoughts. It is the ideas of the ancients that have traveled through time, and it is the echoes of time past that we need to unravel.

In terms of everyday life, getting up in the morning, going to work, and completing household chores, it matters not how the Earth is put together and how the Earth is placed within the cosmos. Even the great Sherlock Holmes could chide Watson (in *A Study in Scarlet*) with the observation that "what the deuce is it to me? [...] you say that we go around the Sun. If we went round the Moon it would not make a pennyworth of difference to me or my work" — and indeed, Holmes is exactly right. For all this, however, we cannot but notice and indeed marvel at the observable heavens, and ultimately, for so it seems, it is human nature to question what the great vista reveled during the night time really means; to ask what is out there, and to question how the whole shebang is put together. Remarkably, and there is absolutely no reason that there should be, there are aspects of the universe that appear to be understandable, and it was the ancient Greek philosophers who first realized that human intellect could both construct and comprehend a working model of the cosmos. Nature, the ancient philosophers realized, was not totally capricious and unknowable. Indeed, they realized that the universe could be understood in rational terms.

While the ancient Greek philosophers were more than up to the intellectual challenge of making sense of the universe, the one thing they lacked was data. They had, of course, what they could see, but there was no established framework within which the observations could fit. In the words of Jonathan Swift, "vision is the art of seeing what is invisible to others." In the modern era it is taken for granted that all observations and specifically scientific observations are predicated upon some form of theoretical foundation or framework. Yes, the ancient Greek philosophers could see the stars and planets, and follow the path of the Moon through the constellations, just as anyone can to this very day; but, they lacked the necessary technology to observe and measure the underlying structure. All they had to interpret an observation with were their imaginations and their common-sense expectations. So, in purely humanistic terms, what was it that the ancient Greeks could see with respect to the Sun, Earth and Moon system? Clearly, for everyday human senses indicate that it is so, the Earth was a fixed center, unmoved and unmoving, and everything, the sky, the planets, the Sun and the Moon were in motion around it — and, by default humanity. The Sun clearly passed below the Earth during the night time, moving from the west where it set to the east where it would rise again. How the Sun did this was unclear, since no one had ever seen over the horizon, or at least any distant horizon that might reveal what it did. No matter, of course, since the blanks in knowledge and direct experience could be filled in by gods, spirits, miracles, and of course good old common sense. As to the shape of the Earth, everyone knew that if you traveled far enough in any direction, north, south, east or west, one would sooner or later encounter the sea, so obviously the Earth, the solid immovable ground beneath our feet, must be supported by a vast all-encompassing ocean. Whether the ocean was endless, stretching forever in all directions, was another matter, and the fact that the Earth was floating on it and the Sun, Moon and stars could apparently survive passage through it, as they rose and set with respect to Earth's horizon, well, this all added to the mystery of creation. This is a complete and perfectly reasonable explanation (by human standards): it fits the known and observable facts and it places the Earth, and us, at the center of the universe — exactly where our collective human egos would expect it to be.

An Earth, of indeterminate and unspecified size, floating in an infinite sea, Oceanus, is essentially the cosmological model that is found in the epic Homeric poems of *The Iliad* and *The Odyssey*. Indeed, this is the universe described and encoded within the design of Achilles' shield. The epics of Homer are thought to have been written some time about the 10th century B.C. They were old even to the ancient Greeks. The picture of a disk-like Earth floating on Oceanus was essentially that adopted by Thales in the mid-5th century B.C. Although there are no surviving works by Thales, he is often credited with being the first of the great Greek philosophers, arguing that there must be natural explanations to physical phenomena and that the universe could be understood according to rational principles. To Thales the idea that the Earth might be floating on a large, if not infinite sea, made (common) sense, and he expanded upon the concept to account for earthquakes (a common enough occurrence in the eastern Mediterranean) — the ground would obviously quake and shudder, he reasoned, as it flexed and rolled in response to the rippling ocean swell. So taken, in fact, was Thales by the concept of Oceanus that ultimately, he argued, everything, deep-down, was made of water — since clearly, as everyone knew, water had the strange ability to be a solid, a liquid and a gas. The cosmological picture attributed to Thales was essentially that developed by earlier philosophers in Babylon and Egypt. It is sometimes described as the four-poster bed model, in that four pillars are envisioned as extended upwards from Oceanus to support the solid heavens (the canopy of the bed) overlapping a flat

Earth (the bed itself). The idea of a solid heaven seems odd to us today, but it was attested to by the fall of meteorites — literally solid pieces of rock and/or iron which fall from the sky. It is a good common-sense association, but, of course, it is entirely wrong. Indeed, the four-poster bed model did not survive for very long once a new breed of critical thinking Greek philosophers emerged on the scene.

By the beginning of the 4th century B.C. the idea of a universal ocean was gone, replaced instead by empty space. Anaximander argued that the Earth was drum-like in shape; 1/3rd as deep as it was wide, and it hovered in space, neither rising nor falling, nor drifting to the left or the right. The Earth was held in stasis, and like Buridan's hapless donkey it had no inclination to move one way or another — supported, as it were, by the power of a philosophical paradox. Xenophanes, unhappy with a dithering Earth, turned Anaximander's drum into an infinitely long cylinder. Humanity lived on one of the exposed ends of the cylinder (a very special location indeed) while its roots sank downwards forever. Horrified at the thought of infinity, Anaximes truncated Xenophanes' infinite cylinder to a flat plate, with a raised rim, being held aloft, like some floating leaf, by a supporting column of air. For Pythagoras in the early 4th century B.C. it was obvious that the Earth could have only one shape — it must be a sphere. For indeed, the sphere, of all the shapes that could be imagined, was perfect and uniform, with no edges nor corners or ugly dimples or bulges. The reasoning was obvious to Pythagoras, since he and his followers believed in a world designed according to numbers and geometry. With nothing more than mathematical aesthetics as their guide, the Greek philosophers latched onto the Pythagorean sphere. Empedocles, in spite of his universal principles of love and strife, order and disorder, accepted that the Earth was a sphere. And, bringing us to the 3rd century B.C., so too did Philolaus of Croton, Eratosthenes, Plato and Aristotle accept the idea that the Earth must be a sphere. For all of the persuasive arguments of these philosophical heavyweights, however, there were dissenters. Anaxagoras, in the late 4th century B.C., flatly rejected the idea that the Earth could be a sphere, and argued instead, just as Anaximes had done earlier, for a shield-like form. Leucippus, founder of the atomist philosophy in the early 3rd century B.C., argued that the Earth was shaped like a snare drum, while his follower Democritus suggested that the Earth must be like an inverted shallow funnel, sloping outwards as one moved towards the south.

Just as the ancient Greeks had a veritable smörgåsbord of shapes and models to choose from with respect to the form of the Earth, so too they also had the literal pick of any number of models for the cosmos. For Anaxagoras, the Sun was a hot, mountain-sized rock located just beyond the Moon — the Moon, however, had no light of its own but was instead empowered by the Sun to shine; for Leucippus the Sun was further away than the most distant planet, Saturn, and adrift in an infinite sea of atoms and stars. To Heracleitus the Sun was just one-foot across, a burning mass of vapor within the Earth's atmosphere, which upon being rekindled simply rose and set every new day. For Anaximander the Sun was a distant window set within a vast tube of fire. To Hipparchos the Sun was located at the very center of the cosmos — to everyone else, until the time of Copernicus in the 16th century A.D., the Earth occupied that exulted location. To the Pythagoreans the Sun was just another planet — its orb, like that of the Earth, set in motion about the all invigorating central fire.

That the Moon must move about the Earth and be closer to the Earth than the Sun was understood by Thales and all other ancient philosophers through observations of the Moon phases and the occasional appearance of lunar and solar eclipses. Indeed, the explanation of lunar phases articulated by Anaxagoras is predicated upon the Moon being a sphere, and eclipse observations indicated that the Moon must be at least ½ the size of the Earth (an overestimate, in fact, by a factor of 2). Any idea of the scale of the Earth–Moon–Sun system, however, was unknown until the work of Aristarchus in the 3rd century B.C. From a brilliant piece of geometrical reasoning, Aristarchus deduced from a measurement of the Moon–Sun angle at the time of a quarter-phase illumination that the Sun must be at least 20 times further away from the Earth than the Moon (see Figure 10.1). Likewise by observing the characteristics of lunar eclipses, Aristarchus estimated that the Moon must be located some 65 Earth radii away from the Earth. Other observations reveal that the Sun and the Moon have about the same angular diameters in the sky — some 0.5 degrees each. This fortuitous circumstance is responsible for the fact that the zone of totality, as experienced during a total solar eclipse, is just a few hundred kilometers across at the Earth's surface. To Aristarchus it additionally indicated that the

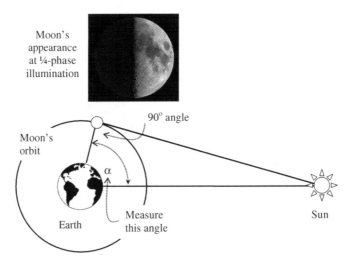

Figure 10.1. Aristarchus estimated that the angle α between the Sun and the Moon when at its first-quarter phase, as seen from the Earth, was not smaller than 87 degrees, indicating that the Sun is at least $1/\cos(87°) = 19.1$ times further away from the Earth than the Moon.

Sun must be at least 20 times larger than the Moon. While it transpires that Aristarchus was wrong by a factor of about 20 in his estimate for the distance to the Sun (it is some 400 times further away than the Moon), his estimate for the Moon's distance was very good. For all this, however, it was a relative distance based upon the unknown size of the Earth — to determine the distance in common units, the stade (a unit to be discussed below) for example, would require a measurement of the Earth's circumference, which in turn would reveal its physical radius. This is where Eratosthenes will eventually enter into our narrative.

Aristotle presented three proofs that the Earth must be a sphere. The first two were based upon actual observations while the third was philosophical. The first proof was predicated on the observation that the shadow of the Earth, as it swept across the Moon's surface during a lunar eclipse, was always curved and only a sphere would produce this effect. The second proof required a trip to the seaside and relied upon the observation that ships disappeared across the horizon hull-down — the very tips of their top sails being the last vestiges of their rigging to pass from the observer's view. Both of these observations are sound and they point towards the Earth being at least spheroidal in shape. The third proof presented by Aristotle was related to his doctrine of final causes. Light-weight substances, such as smoke and vapors, he reasoned, tended towards the limits of the cosmos, and therefore, if unconstrained, would rise upwards from Earth's surface. Heavy objects, in contrast, tended towards the center of the cosmos, which was their natural place to reside. Heavy objects fell towards the Earth's center, in Aristotle's philosophy, simply because the Earth, made as it was of heavy material, surrounded the center of the cosmos. The subtlety here is that the Earth is not the center of the cosmos, but is rather located at the center of the cosmos because of what it is made. That the Earth must accordingly be a sphere then comes about because all of its constituent parts are trying to get as close as they possibly can to the center of the cosmos, and the figure that accommodates the maximum volume with a minimum surface area is that of a sphere. Aristotle presented a number of estimates for the size of the Earth, but he offers no indication as to how the estimates were made. As to whether they were pure guesswork or based upon some form of actual measurements we do not know, but the value presented by Aristotle in his *de Caelo* (written circa 350 B.C.) is 400,000 stadia. The other great mathematical philosopher Archimedes likewise provides an unexplained estimate of 300,000 stadia for the Earth's circumference in his *Sand Reckoner* (published circa 250 B.C.). It is not until a century after the death of Aristotle that we first hear of the Earth being physically measured and then it is only through an account that was written some 500 years after the event took place. The key player in this early measurement was Eratosthenes, and while we do not have his original, or even copies of his original works, his achievements have echoed down the centuries to be celebrated even to the present day.

Eratosthenes was born in Cyrene (in modern day Libya) around 285 B.C. We know nothing of his early life, other than he eventually moved to Athens to study at the school established by Zeno of Citium. The Roman historian Pliny the Elder, nearly 200 years after Eratosthenes had died, wrote that he was a "man who was peculiarly well skilled in all the more subtle parts of learning." Indeed, Eratosthenes clearly made a name for himself in Athens since in 245 B.C. he was engaged by Ptolemy III

(Euergetes — *the Benefactor*) to be Head Librarian at the great library and museum in Alexandria. At the time of this life-changing move Eratosthenes would have been about 40 years old. Founded circa 300 B.C. by Ptolemy I (Soter — *the Savior*), the library at Alexandria was to become the greatest repository of knowledge in the ancient world, and it survived essentially intact until the final sacking of Alexandria by Aurelian in 270 A.D.

At Alexandria Eratosthenes apparently excelled; producing there a whole series of new and important texts on geography, astronomy, and mathematics. Cleomedes, writing circa 350 A.D., indicates that Eratosthenes made his famous deduction about the size of the Earth in 240 B.C. At later times in Alexandria he is credited with the invention of the armillary sphere, an analog model of the heavens, and a mathematical sieve technique for finding prime numbers. His works relating to ethnology, topography and geodesy earned Eratosthenes, in later centuries at least, the title of *father of geography*. Indeed, at Alexandria, Eratosthenes's studies roamed both wide and deep and he adopted the name *Philologos*, which translates to something like a lover of *logos* — with *logos* essentially relating to logic, words and proportion.

Exactly what set of circumstances led Eratosthenes to make his determination of the Earth's circumference is unknown, but they must have involved at least one fortuitous tidbit of information. While it is known that Eratosthenes did not travel from Alexandria, he nonetheless discovered that in the town of Syene, located on the river Nile in southern Egypt, at noon on the day of the summer solstice, the Sun shone directly down a deep well. On any other day of the year, the Sun would only illuminate the sides of the well at Syene. This, as Eratosthenes realized, was a significant observation, since on that same day and at the same time in Alexandria the Sun was not directly overhead — indeed, it was never overhead in Alexandria. Measurements (presumably by Eratosthenes) revealed that at noon on the day of the summer solstice the Sun was 1/50th of a full circle (or about 7 degrees) away from the zenith. With this information in place it is possible to set about estimating the size of the Earth. To perform this task, however, one additional measurement needs to be secured and two assumptions have to be made. The two assumptions relate to the geometry of the situation to be analyzed (Figure 10.2). The first assumption requires that sunrays, when they strike the Earth's surface, run parallel to

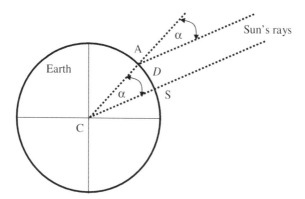

Figure 10.2. Eratosthenes's method for finding the Earth's circumference. A and S correspond to the locations of Alexandria and Syene respectively, *D* is the distance between Alexandria and Syene and α is the zenith angle of the Sun from Alexandria at noon on the day of the summer solstice.

each other, and this in turn requires that the Sun be located well away from the Earth. The second assumption in the analysis is that the Earth is actually a sphere. Eratosthenes did not prove that the Earth was a sphere — rather he determined the size of the Earth on the basis that it is a sphere. And, finally, Eratosthenes needed an estimate for the distance *D* between Alexandria and Syene. Many suggestions have been made as to how Eratosthenes might have made this latter measurement, from paying someone to pace out the distance on foot, to estimating the speed and travel time of a boat, along the Nile, from Syene to Alexandria. The actual method Eratosthenes used is more practical and less romantic — and it did not require any new measurements to be made. His distance of 5,000 stadia between Alexandria and Syene was based upon land measurements made for tax collection purposes, and this was data readily available to a scholar with the right connections in Alexandria. With this final piece of information in place, the distance around the Earth's circumference *C* can be determined. It is simple geometry (Figure 10.2) that gives the final result that $D = 5,000$ stadia $= (7/360)C$, or after a little rearranging, the Earth's circumference $C = 250,000$ stadia. Knowing additionally that the circumference of a circle is equal to 2π times its radius R (that is $C = 2\pi R$), so the Earth's radius turns out to be about 40,000 stadia [21].

It is at this stage that we hit a near insurmountable problem. The units of measure for the distance between Alexandria and Syene is given by Eratosthenes in stadia, but it is not known which stadia unit he was actually using — it was not a standardized distance. Being literally based upon the length of a sports stadium, many different values for the stade are known. Various conversion estimates have been made by classicists and they vary from 1 stade being as small as 157 meters (the Itinerary stade) to as large as 209 meters (the Phoenician–Egyptian stade). It is generally assumed that Eratosthenes used the Attic stade, which is taken as being equivalent to 185 meters. Given the methods available to Eratosthenes it seems somewhat unfair to compare his results with those of the modern era, but on the basis that he used the Attic stade then his value for the Earth's radius is overestimated by some 15.5%. If we allow for the whole range of known stade equivalences, then Eratosthenes's value for the Earth's radius falls within the range $6{,}247 < R_{Earth} \text{ (km)} < 8{,}316$ — so, anywhere from a 2% underestimate to a 30% overestimate compared to the modern value. Irrespective of the actual conversion factor that should be applied, Eratosthenes obtained a remarkable result, and he most certainly improved upon the apparent guesswork (both being overestimates for the size of the Earth) provided by Aristotle and Archimedes.

With the measurement by Eratosthenes for the size of Earth being made, the proverbial cat was out of the bag, and subsequent cartographers either adopted his results or set out to improve upon them. Again, we have only anecdotal results to work with, but it would appear that the next major effort, according to Cleomedes writing circa 150 A.D., was that by Posidonius in about 100 B.C. In principle Posidonius outlined a more refined methodology than that of Eratosthenes for finding the size of the Earth — in practice, however, Posidonius was less exact in his measurements. As with the method developed by Eratosthenes, key to making the determination of the size of the Earth is a measure of the distance, and the difference in the angle of latitude, between two well-separated locations. Posidonius used observations set on the island of Rhodes in the north and Alexandria in the south. He had no good measure for the distance between these two locations, which spanned the divide of the Mediterranean, but he took it to be 5,000 stadia (the same

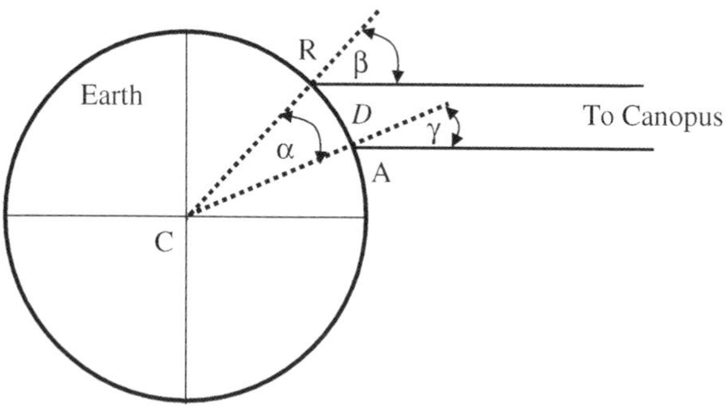

Figure 10.3.   Posidonius's method for determining the Earth's circumference. R and A correspond to the locations of Rhodes and Alexandria. As illustrated here the method of Posidonius is to measure the zenith angles β and γ, with straightforward geometry then giving $\alpha = \beta - \gamma$. Posidinous made use of the observation that from Rhodes the angle β is actually close to 90° — that is, Canopus barely ascends above the horizon from that particular location.

separation that Eratosthenes had deduced from the land taxation surveys between Alexandria and Syene). This distance, from what can be deduced from the unit of measure, is in fact an overestimate of the separation by about a factor of 5/4. To determine the difference in latitude between Rhodes and Alexandria, Posidonius relied upon observations of the bright star Canopus (Figure 10.3). By using a star (rather than the shadow length measurement used by Eratosthenes) as the reference object, Posidonius could in principle measure the difference in latitude between Rhodes and Alexandria with great accuracy. This result comes about because a star is a point source, in contrast to the Sun which is an extended object in the sky. Therefore, if the altitude of Canopus is simultaneously measured from Rhodes and Alexandria this would provide an accurate measure of the latitude difference. Unfortunately, rather than actually make any refined measurements, Posidonius chose to use hearsay about the appearance of Canopus instead. From Rhodes, Posidonius argued Canopus barely rises above the horizon, and then only briefly for a few days per year. In Alexandria, however, Canopus reaches a maximum altitude of 1/4 of a sign (that is 1/48th of the zodiac, or 7.5 degrees) above the horizon. Neither of these results is actually correct, but they come out to be close to the correct angle required for a separation of 5,000 stadia as determined by Eratosthenes. Accordingly, Posidonius determined that the Earth's circumference amounted to a distance of 240,000 stadia, giving a radius of 38,000 stadia — just a few percent different to the radius determined by Eratosthenes. While Posidonius made no actual observations, he was fully aware of the fact that an accurate distance measure between the two locations at which the angular heights for Canopus were measured was crucial to the end result. For a given variation in altitude difference, the smaller the distance between the two observing locations, the larger must the Earth be and *vice versa*. What history and commentaries give us next is confusion. The Greek geographer Strabo, writing circa 50 B.C., records that Posidonius calculated the circumference of the Earth to be 180,000 stadia — a value at great odds and significantly smaller with that derived by Eratosthenes. Of all the commentators that mention Posidonius's work, it is only Strabo who mentions this small circumference (a value that implies an Earth radius of some 28,650 stadia, or about 4,584 km), and yet it is Strabo's singular comment that has echoed through history. The smaller value attributed by Strabo to Posidonius for the Earth's circumference cannot be explained by known variations in the stade unit of measure, and in reality the number is probably just a typo or mistranslation somewhere in the line of copies of Strabo's work that have come down to us. For all this, however, it was partly upon the smaller Earth circumference (supposedly) set by Posidonius that Christopher Columbus, in the 15th century A.D., justified his campaign to sail westward from Spain in order to reach a relatively close India in just a matter of days[1] — what he discovered, of course, was the North American continent instead. Claudius Ptolemy, writing in the first century A.D., without referencing Posidonius, additionally attributed a small size for the Earth, but in Ptolemy's case he simply states the units that he is going to use on his maps: 500 stadia to 1 degree of latitude or longitude. For map making this assumed conversion is fine on a local scale, but it would ultimately result in a global mismatch, with the paced-out stadia measure, assuming it could be made, revealing more than 360 degrees around the Earth's circumference.

---

[1] This and the fact that Columbus additionally used an incorrect (underestimate) conversion of one degree of latitude into miles.

In terms of shear brilliance and actual measurements, the first accurate determination of the Earth's radius was made by Iranian scholar Abu Rayhan Biruni (generally known to European scholars as al-Biruni) circa 1000 A.D. Al-Biruni spent much of his life in Ghazni in modern-day Afghanistan and his method for finding the size of the Earth was inspired by looking up and measuring the height of a near-by mountain. The key and innovative point with respect to al-Biruni's methodology was that it did not require any distances to be paced out, and all the observations were made local to the mountain under study — it does require the observer to climb the mountain, however. To find the height *h* of the mountain al-Biruni used the method of similar triangles and a square-shaped measuring instrument (Figure 10.4). With the height of the mountain determined, the observer must then climb to the top of the mountain and measure what is now called the angle of dip (Figure 10.5). The angle of dip φ is simply the angle between the horizontal and the horizon as seen from the mountain top. With the angle φ measured, so (in modern terminology) $\cos \varphi = R/(R + h)$ where *R* is the radius of the Earth. Al-Biruni determined highly accurate values for *h* and φ, with (again using modern units) the height of the mountain being 305.1 meters and the angle of dip being 0.57 degrees. These numbers give an Earth of radius $R = 6{,}238$ km, which is just 2% smaller than the modern-day reference value. The method developed by al-Biruni is simply beautiful and beautifully simple. It does away with the problem of determining

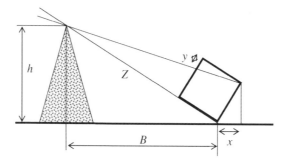

Figure 10.4. To determine the height *h* of a mountain, align the base of the square frame (of side lengths *L*) with the mountain top, and then determine the distances *y* and *x*. By similar triangles, $Z/L = L/y$, which gives *Z*, and $h/Z = x/L$, which gives the height of the mountain *h*.

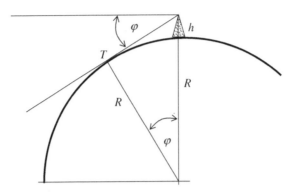

Figure 10.5. Determining Earth's radius. From the top of a mountain of height h, measure the angle φ between the local horizon and the tangent point T. The radius of the Earth is then given by the relationship: $\cos \varphi = R/(R + h)$.

the distance between two well-separated observation sites and there is no need to adjust the observations for any off-set in longitude as is required in the method of Eratosthenes and Posidonius.

Ptolemy's production of his *Mathematica Syntaxis*, more generally known as the *Almagest* after its later Arabic translational title, in the first century A.D. marks the high point of ancient Greek astronomy. The *Almagest* was to become THE astronomy book for the next 1,400 years, and it contained all that one needed to know about predicting the heavenly motion of the Sun, Moon and planets. It was a summary of genius and a work of genius. Perhaps, one could argue that the *Almagest* was too good, far too good in fact, since while it was the focus of many reviews and commentaries, especially by Arabic scholars, no reasonable alternatives could be found to replace its planetary epicycles, eccentrics and deferents. All this aside, we nonetheless have by the time that Ptolemy was writing, a highly sophisticated set of geometrical models to describe planetary motion within a cosmos centered upon a spherical Earth — and the size of the central spherical Earth had been reasonably well determined through direct experimental measure.

With the establishment of universities in most major European cities, from the 11th century onwards a new interest in the scholarly investigation of the heavens began, and one of the most successful and widely read texts to be produced was that by Johannes de Sacrobosco (John of

Hollywood). Based at the University of Paris, Sacrobosco produced in circa 1230 a short review of astronomy under the title *Tractus de Sphaera* (*On the Sphere of the World*, where in this case *World* means the heavens rather than the Earth itself). This tome was heavily based upon Ptolemy's *Almagest*, but it essentially became required reading for any student undergoing a university education for the next several hundred years. Although primarily about the heavens, Sacrobosco's *Sphaera* does contain many detailed accounts about the Earth. "The Earth is a sphere," writes Sacrobosco, "with a total girth [...] comprising 252,000 stadia." This number for the Earth's circumference is based upon the measure of 700 stadia for each of the 360 degrees around the equator. As proof of this statement Sacrobosco argues that the measured change in the altitude of the north celestial pole (NCP) varies by one degree for every 700 stadia one moves northward in latitude. Here Sacrobosco is placing the Earth at the center of a spherical cosmos — the celestial sphere of the ancient Greeks. The mathematical proof that the altitude (that is angle above the observer's horizon) of the NCP corresponds to the observer's latitude (that is the angle above the equator) is straightforward and illustrated in Figure 10.6. The proof is based upon definitions and by

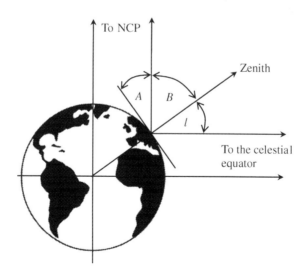

Figure 10.6. The determination of latitude by measuring the altitude of the north celestial pole. Angle *A* corresponds to the measured altitude of the NCP; and *l* is the observer's latitude upon the Earth's surface.

accounting for the sum of various angles. Suppose the observer is at some latitude *l* above the equator, and draw a line from the center of the Earth through their feet and continue this line outwards to the observer's zenith *Z* on the celestial sphere. Now, by definition, the observer's horizon is that plane (line in the diagram) extending from the observer's feet at 90 degrees to the observer's zenith. The altitude of the NCP is angle *A*, and this angle can be measured. Once again, by definition, we know that angles *A* and *B* must sum to 90 degrees. We can also see by construction that angles *l* and *B* must also sum to 90 degrees. A straightforward bit of algebra now gives the result that *l* = *A*, that is, the altitude of the north celestial pole above an observer's horizon corresponds to the latitude of that observer on Earth's surface. This result, of course, is highly useful for navigation purposes in the sense that it allows for the determination of one's location on the Earth (north of the equator) via a relatively straight-forward measure of the altitude of a specific sky location — recall, as every Boy Scout knows, that the North Star (Polaris = α *Ursae Minoris*) chances to reside at the location of the NCP at the present epoch. Once again, Sacrobosco is using the somewhat problematic unit of stade to describe the circumference of the Earth, but on the principle that there are 700 of them per degree around the equator, then in modern terms 1 stade corresponds to a distance of about 159 meters.

Through Sacrobsoco we see the presentation of a very practical astron-omy, and his emphasis is directed towards the utilitarian aspects of astro-nomical measurements and their use in organizing life, commerce and navigation on Earth. At this stage in history there is no real concept of astronomy as being the study of distinct objects beyond the Earth's atmos-phere. Indeed, at this time (the 12th and 13th century) there are no con-cepts, other than their observed motion in the sky, of what makes a planet different from a star, and likewise there is no clear concept as to whether the Sun is a star, or that the stars are suns. Not only is there no sense, in the early medieval period, of astronomy being the study of distinct objects in the cosmos (as in the modern era), there were likewise no clear ideas about how motion *worked* upon the Earth's surface. This latter topic, however, began to see a gradual drift away from the ideas of Aristotle in the mid-14th century when the, so-called, Merton calculators at Oxford University (see the next chapter) began to study anew the interrelationships between displacement, velocity, change in velocity (acceleration) and time. Indeed, it is not until the mid-14th century that the concept of time, as a quantifiable phenomenon, is introduced into natural philosophy. Prior to this development, time was viewed as being something that was entirely subjective, and under such conditions it is difficult, nay impossible, to develop a clear concept of speed and motion — speed after all is defined as being the distance traveled per unit time (but more on this topic later).

The first practical demonstration that the Earth was spherical was made in the 16th century through the expedition organized by Ferdinand Magellan. Intent on finding a new shipping route to the Spice Islands, and of course profiting from the discovery, Magellan set out from Spain in 1519 with a fleet of 3 ships manned by 237 sailors and crew. Three years later, after circumnavigating the globe, the remnants of Magellan's fleet returned to Spain. Only 18 of the original crew, Magellan not one of them, survived the rigors of the entire journey, reflecting a mortality level of 92 percent. Early exploration of the world's oceans was not only slow, it was additionally highly dangerous. Within two decades of the return of Magellan's depleted fleet to Spain, the now circumnavigated Earth was to undergo further human-inflicted upheaval. Not only had humankind finally assailed its girth, but in 1543 human ingenuity displaced it away from the center of the cosmos.

The great work of Nicolaus Copernicus, *De Revolutionibus Orbium Coelestium*, appeared in print form as he lay on his deathbed, but it had been a work that had occupied his mind for the previous 30 years. The book was an outright attack on Ptolemy's *Almagest* and the way that astronomy had been conducted during the previous one-and-a-half millennia. Copernicus set out to change the dynamics of the heavenly bodies, and in this quest found that the best place to locate the Earth was as the third planet out from the Sun. The Copernican revolution, as it is usually described, was not a necessary revolution, there was no conflict between the predictions of the Ptolemaic theory and the observations (that couldn't be readily accounted for). Rather, the revolution was in thinking, and indeed, Copernicus essentially presented the results of a complex thought experiment. His approach was one of mathematical simplification guided by esthetics. How, he reasoned, could the heavenly motion of the planets be described in terms other than the ancient Greek epicycles and deferents. One foot in the past and one in the future, Copernicus set the Sun at the center of a still finite, but now stationary celestial sphere (which contained all the stars), and put the planets in circular orbits around the central "glorious lamp." Scholars have long argued whether Copernicus really believed in the physical structure of his model, with the planets actually moving around the central Sun, or whether it was intended as a purely mathematical device for calculating planetary positions. That Copernicus did not wish to publish the details of his revolutionary tome during his lifetime is entirely understandable — he knew that its printing would upset just about everybody. Not only do we find poet John Donne bemoaning the fact that the "new philosophy calls all in doubt [...] Tis all in pieces, all coherence gone, all just supply, and all relation"; but so too did the three pillars of the establishment tremble and moan at the implications of the Copernican theory. These pillars of Aristotelian authority, Church authority and human bravado, were all insulted by the Copernicus model: humanity, invoking a bruised ego, was displaced away from the center of the universe, and Church authority was threatened since it was then built around the Aristotelian doctrine of a central Earth. As a predictive device for determining planetary positions, the Copernican model, in its first form, was less accurate than that of the Ptolemaic epicycles, but its simplicity of design was the key feature that caught the eye of the new luminaries, Galileo Galilei and Johannes Kepler. It was these two players that in the early 17th century picked up the Copernican idea and made it work. Indeed, these new explorers were revolutionaries in their own right, and it was they who not only fine-tuned the workings of the Copernican model, but also forced the issue with respect to the fact that it represented physical reality — the Earth really was the third planet out from the Sun.

Kepler was the mystic mathematician, while Galileo was the promotion-hungry physicist. It was Kepler, after his many years of "war with Mars," who established the elliptical nature of planetary orbits and through his three laws of planetary motion established the rules by which the planets are marshaled with respect to their paths about the Sun. It was Galileo who hammered away at the Aristotelian doctrine, arguing that the wisdom of the ancients should not be trusted, and that new and independent methods of inquiry should be used to determine how nature and the universe really operated. While Galileo vociferously promoted the Copernican model, and is often seen as a martyr to its cause, he did nothing to correct its known inadequacies and in spite of his bravado he could not prove (even beyond any reasonable doubt) that the model was a representation of reality — all such proofs came along much later, long after Galileo's death. For all this, however, Galileo was integral in advancing the study of terrestrial motion, and indeed his *Dialogue Concerning Two New Sciences*, published in 1638, is an absolute belter! This text, as we shall see later, was pivotal in pining down key experimental aspects to accelerated and constrained motion. Indeed, Galileo described in detail the motion of a pendulum (getting most of it right), and he also opened up the study of kinematics under the condition of constant acceleration.

Neither Copernicus, nor Kepler nor Galileo had much to say about the Earth itself. That it was a sphere was taken as proven, and for their purposes its physical size was not of specific importance. Other practitioners in the late 16th century, however, felt that it was time to begin anew, with modern methods and instruments, the measure of the Earth. The outlines of how land might be accurately surveyed, through geometrical triangulation, were first outlined by Gemma Frisius in his 1533 corrected and expanded version of Petrus Apianus's *Cosmographia* (which was first published by Apianus in 1524). *Cosmographia* was essentially a layman's guide to all matters practical with respect to astronomy, geography, surveying and navigation, and in his expanded edition Frisius introduced the reader to the triangulation method that is still used by surveyors to this very day. As a mathematical concept, the process of triangulation is straightforward and it only requires that one side of the first triangle be physically paced out on the ground. After the baseline has been determined, the remainder of the observations relate to the determination of angles. While straightforward on paper, triangulation is fiendishly difficult to put into practice, as astronomer Tycho Brahe and his assistants discovered when they tried, between 1588 and 1591, to put Frisius's ideas into practice in the mapping out of the island of Hven. Dutch astronomer, mathematician, and former member of Brahe's *famalus*, Willebroad Snellius[2] was the first practitioner to attempt a measure of the Earth's circumference by mapping out a one-degree arc of longitude by triangulation in 1615. Snellius found that such an arc corresponded to a distance of some 28,500

[2] Probably best remembered today through Snell's Law of refraction.

Rhineland Rods, or in modern units, about 107.4 kilometers. Such a measure implies an Earth radius of 6,152 km, which is some 3.5 percent smaller than the modern-day standard. Between 1644 and 1656 Giovanni Battista Riccioli and Francesco Grimaldi attempted an improvement on Snellius's measure, and concluded that 1 degree of latitude corresponded to a distance of 373,000 Pedes (a Roman unit of about 29.5 cm in length). With the improved instrumentation available, the Earth radius determined by Riccoli and Grimaldi came in at about 6,304.5 km, which is within about 1 percent of the modern-day value.

Rather than rely upon triangulation techniques, British mathematician and surveyor Richard Norwood[3] decided to measure the length of one degree of arc courtesy of the practicality of the Romans. To this end, between June 1633 and June 1635, he set out to both measure the latitudes corresponding to London and the City of York and to pace out the distance between them. His methodology was going to be exactly that outlined by Eratosthenes, and the link with the ancient Romans was that the great Ermine Street, built between 45 and 75 A.D., ran pretty much in a straight line due north, along the backbone of England, from Bishopsgate in London (*Londinium*) through *Lindum Colonia* (Lincoln) and on to *Eberacum* (York). Using both chain and pacing methods Norwood recorded over a "ten or eleven days" walk between London and York that the distance between the two centers was 9,149 chain lengths. Working these units to something more recognizable in the modern era, it transpires that 1 chain corresponds to 6 Poles, and that 1 Pole is equivalent to 16.5 feet — and this gives us the distance between London and York as being 905,751 feet (or 276.1 kilometers). Furthermore, Norwood obtained from noontime Sun-altitude measurements that the City of York was 2 degrees 28 minutes further north in latitude than London. Publishing his results in 1644 in a relatively short text with the long title of *The Seaman's Practice: containing a fundamental problem in navigation experimentally verified*, Norwood deduced that 1 degree of latitude corresponded to a distance of 367,196.35 feet, or about 69.54 miles (111.9 km). With this number quantified, Norwood accordingly revealed that the Earth, taken as a perfect sphere, has a circumference of 25,034.4 miles (40,287.1 km) and a corresponding radius of some 3984.3 miles (6411.9 km). Norwood's result is

---

[3] Norwood is additionally remembered for his extensive surveys of Bermuda conducted in 1613 and 1662.

within 1% of the modern-day (average) Earth radius — not a bad result at all. Indeed, none other than the esteemed Isaac Newton adopted Norwood's value for the Earth's radius in his *Principia*, although Newton also introduced a complication, in that he showed that the Earth cannot be a perfect sphere — the Earth, in fact, must be an oblate spheroid; the Earth's equatorial radius being appreciable larger than its polar radius. We shall examine Newton's reasoning for postulating a non-spherical Earth in Chapter 14, and simply note here that the observations from the 19th century onwards have indicated an ellipticity of about $e \approx 1/300$, where the ellipticity is defined as being $e = (a - b)/a$, where b is the length of Earth's semi-minor axis and a the length of its semi-major axis. Present-day values indicate that Earth's polar radius is about 27 km shorter than its equatorial radius. Like the profile of old professors, the Earth bulges outward when viewed across the middle. It is because of this equatorial bulge that the farthest one can stand away from the Earth's center is not the summit of Chomolungma (Mount Everest) in Tibet, but the summit of Mount Chimborazo in Equador (Figure 10.7).

Figure 10.7. *Study of Mount Chimborazo*, Ecuador, by Frederic Church (1857). The peak of Mount Chimborazo (4,118 m) represents the greatest distance that one can obtain from Earth's center. Now an inactive stratovolcano, Chimborazo last erupted some 2,500 years ago.

Chapter 11

Aristotle's Stop and the Merton Calculators

How do you expect to arrive at the end of your journey if you take the road to another man's city?

New Seeds of Contemplation, Thomas Merton (1962).

To Aristotle's way of thinking, the world, and specifically the physical world, was something that could at best be qualitatively understood. It was not, as he saw it, something that was knowable through quantitative analysis. Indeed, he openly criticized those who would reduce the study of the world to mathematics. In this sense Aristotle rejected the concepts outlined by his tutor, Plato, that mathematical objects really do exist, and, indeed, have an existence independent of the human mind. While Plato was quite happy to accept the notion that lines could be infinitely long, or that certain sets could contain an infinite number of objects, Aristotle would have none of it — to Aristotle there was no room for the infinite in the human world, which he considered to be very much a finite domain. Indeed, as was described in Chapter 10, Aristotle offered three proofs (as if one was not enough) that the Earth must be a sphere of finite size. Aristotle's third proof for a spherical Earth involved his ideas concerning purpose — a concept that not only infused within all objects an innate capacity for change, but also dictated the type of change that a given object might experience. Earthy, solid objects fell downwards once released from some elevation because it was their in-built purpose to strive towards an *end* that would place them at the center of the cosmos — the Earth, recall, then must be a sphere since this is the tightest packing of all its earthy material all jostling and striving to get as close

119

to the center of the cosmos as possible. To Aristotle it was clear that objects could move, and some objects moved further in a given time interval than others, but it was always a qualitative motion with concepts such as speed and acceleration having no clear meaning at that time. Indeed, to the ancient Greek philosophers, time was part of the problem in the sense that they had no way of measuring it precisely, and accordingly it would have been pointless (even if the idea had crossed their minds to do so) to perform quantitative experiments on how objects moved and how their speed might change as they fell. For all this, however, Aristotle did posit statements about motion, arguing, for example, that massive objects fall more rapidly than lighter ones — a *fact* that is born out by human experience, but not an actual fact of real motion.

It was not until the 14th century that Aristotle's ideas concerning the purpose of motion were meaningfully challenged and that new ideas concerning the meaning of velocity and acceleration were developed and explored. And, even then, it was not an experiment-based exploration that took place, but rather it was an exploration in terms of thought experiments. Certainly, Latin translations of Aristotle's works, from Greek and Arabic sources, had been available to the medieval scholar from the late 12th century onwards, but it took several centuries for their meaning and content to be unpacked, appreciated and understood. Theologian St. Thomas Aquinas (c. 1235–1274) began the process of assimilating Aristotelian principles concerning the order of nature into Church doctrine, and indeed used such principles to argue for the existence of God. Concerning motion, for example, Aquinas writes in his *Summa Theologiae* (Part 1, question 2, articles 3 — written circa 1270) that "it is evident that they reach the end not by chance but by intention. Thus, there is something intelligent, by which all natural things are ordered to an end: and we call this God." Here we have to be a little careful in the sense that by motion Aristotle and Aquinas were referring to change of every kind, not just that of falling objects, but of such phenomena as weather, birth, life and death, the heavens, and the seasons. For all this, however, Aquinas did begin to question, "why do objects move?" Not only were scholars exploring the nature of motion during the 13th century, but they were also developing the mathematical tools and concepts to describe what was going on. Robert Kilwardby in circa 1250, for example, provided a detailed account in his *De Ortu Scientiarum* of the origins of the sciences (that is knowledge) and the importance of mathematical theory in describing natural phenomena. Kilwardby's student, Roger Bacon raised the mathematical mantra to even greater heights, arguing that "mathematics perfects natural sciences by giving an ultimate explanation of natural phenomena [...] It is impossible to know the things of this world, unless one knows mathematics." With Kilwardby, Bacon and later scholars, we begin to see the establishment of quantifiable mathematics as the bedrock of scientific enquiry. Indeed, mathematics, in the sense outlined by Plato, is derived from its own, real but distinct world — an abstract world of ideals, perfect forms and absolute proofs.

A new mathematical approach to the study of dynamics was initiated by Thomas Bradwardine (c. 1295–1349) in his 1328, *Treatise on the Proportion of Velocities in Moving Bodies*. Bradwardine was a Fellow of Merton College, Oxford (and later Archbishop of Canterbury), and through his works he became known as *Doctor Profundus*. In remarkably modern-sounding tones Bradwardine writes in his *Treatise* that "it is [mathematics] which reveals every genuine truth, for it knows every hidden secret [...] whoever, then, has the effrontery to study physics while neglecting mathematics, should know from the start that he will never make his entry through the portals of wisdom." Strong words, indeed, but words that have echoed meaningfully down the centuries to the modern era. Importantly for our narrative, however, it was in his *Treatise* that Bradwardine finally did away with Aristotle's ideas of motion as a process and the workings of an active potency inherent within all objects, and he specifically spells out the idea that speed is a quantitative phenomenon related to the ratio of a distance traversed divided by the time required to cover the specified distance. Here is the origin of the now known-to-all formula which states: speed = distance/time.

Bradwardine's *Treatise* inspired other scholars at Merton to investigate the mathematical laws of motion. Preeminent among these followers were William Heytesbury, John Dumbleton and Richard Swineshead. It was Heytesbury (c.1313–1373) in particular who developed what has become known as the Merton mean-speed theorem. This rule followed directly from the question "what happens when a moving object accelerates at a constant rate." Heytesbury, like the other Merton scholars,

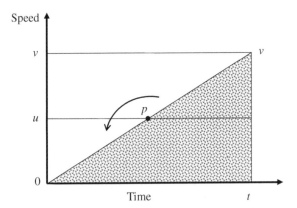

Figure 11.1. The mean-speed theorem. A geometrical proof follows by simply rotating the top part of the shaded triangle about point *p* so as to construct a rectangle having a height given by the mean speed $u = v/2$ and length *t*. The acceleration corresponds to the slope of the triangle's hypotenuse: $a = v/t$.

worked entirely in terms of thought experiments and numbers, but derived, nonetheless, a correct expression for the distance traveled by an object under constant acceleration. Sounding somewhat awkward and decidedly strange to the modern ear, the mean-speed theorem was stated as follows: "A moving body will travel in an equal period of time, a distance exactly equal to that which it would travel if it was moving continuously at its mean speed." Figure 11.1 shows a graphical account of the mean-speed theorem, and indeed, the proof of the statement was presented geometrically. In modern terms we would say that the distance traveled *s* in time *t* is the area under the triangle $0vt$ (shown shaded in Figure 11.1), giving: $s = \frac{1}{2} tv$. If we further denote the acceleration *a* as the ratio $a = v/t$ (the slope of the line in Figure 11.1), then we recover the result that $s = \frac{1}{2} at^2$. This same result is also obtained by considering the area of the rectangle defined by the mean speed, $u = (0 + v)/2 = v/2$, where it is assumed that the object was initially at rest. Accordingly, $s = ut = (v/2)t = (at/2)t = \frac{1}{2} at^2$. It was the mean-speed theorem that paved the way, eventually, for Galileo and Newton to develop their own specific rules and laws of motion in the 17th century.

Richard Swineshead (c. 1340–1355) took 14th century mathematics to its very limits, and is particularly famed for his treatise *Liber*

Calculationum, a study in which he developed a mathematical account of how a heavy object will fall towards the center of the Earth. This, of course, was a fundamental question at that time since the center of the Earth coincided with the center of the cosmos — so the question was really what happens to the motion of an object as it approaches the very center of the universe. Swineshead attacked the problem in terms of forces, with one force $F+$ acting to move an object in one direction and another force $F-$ acting to move it in the opposite direction. If $F+$ and $F-$ are equal then an object remains at rest, but if an object is actually moving then its velocity will be proportional to the ratio $F+/F-$. Swineshead then breaks the problem down by considering the motion of the heavy (extended) body as it approaches the center of the Earth (that is the center of the cosmos) one small spatial interval at a time, with each successive interval being half the size of its predecessor. What Swineshead deduces is that as the center of the falling body approaches the center of the Earth, so its velocity must slow down, but he also finds that the velocity slows down more rapidly than the successive intervals through which the body has to pass. He concludes that the body "will not traverse the whole distance in any finite time." The argument that Swineshead presents to his readers, therefore, is that the falling mass can approach arbitrarily close to the center of the Earth (the cosmos) within a finite amount of time, but that the center of the falling body and the center of the Earth (the cosmos) will never actually coincide. In short, the center of the falling body follows a slow Zeno paradox-like march towards the center of the Earth, but it never actually arrives there [22]. This result has a certain resonance with a question concerning a falling bullet posed by the "Rev. Mr. T-H" in the October 1757 issue of *Martin's Miscellaneous Correspondence* (see the Appendix).

Swineshead's result concerning motion of an object falling towards the center of the Earth is the same result that Aristotle would have derived, but Swineshead has reasoned his conclusions mathematically and according to what were then thought of as reasoned physical principles. That Swineshead's conclusion is actually wrong is not the point at this stage, rather, the point is that an attempt had been made to understand motion in terms of mathematical principles. It was a task left for future scholars to work out the correct physical details. Working at about the same time as the Merton calculators were struggling to understand the theoretical concepts of motion, another Merton scholar, Richard of Wallingford (1292?–1336), was turning his theoretical and mechanical skills to the measurement of time (Figure 11.2). Indeed, if one is to understand what speed is then an understanding of time is essential. Not specifically motivated by the question of dynamics, Wallingford is best

Figure 11.2. Reproduction of a 14th century miniature showing Richard of Wallingford with a pair of measuring compasses. Note the astronomical quadrant to the middle left of the image. The staff indicates that Wallingford was an Abbot at St. Albans, and the facial maculae are indicative of the leprosy from which he suffered.

remembered for the astronomical clock that he described in his *Tractus Horologii Astronomici* (published in 1327 — the same year that Wallingford was elected Abbot at St. Albans Abbey). This great clock and model of the heavens was built within the confines of the Abbey and its construction occupied 20 years of Wallingford's life. Not a clock in the sense that we would recognize today, Wallingford's timepiece was a device that chimed the hours[1] and drove an astronomical display revealing the zodiacal positions of the Sun and Moon — it also indicated the phase of the Moon — and it indicated which of the primary constellations were visible from St. Albans. The Wallingford clock was probably not very accurate, but it was the beginning of a totally new technology which, in later centuries, resulted in the construction of exquisite and highly accurate timepieces. Such later timepieces were central to new developments in natural philosophy and astronomy, and they also (care of John Harrison's H5 marine chronometer constructed circa 1770) enabled safer navigation at sea. Wallingford additionally built and left plans describing the construction of various astronomical devices, literally analog computers, that could be used for measuring star or planet positions in several coordinate systems simultaneously (the so-called *torquetum*), and he made a number of technical improvements to the *equatorium* (a device for predicting planetary positions, which he somewhat mysteriously called the Albion). Wallingford's original clock was destroyed, or at least lost, when King Henry VIII, in 1546, dissolved the Abbeys of England. A replica of the Wallingford clock has more recently been constructed, completed in 1995, and is on display to visitors at St. Albans Cathedral.

---

[1] The word clock is derived from the Latin *clocca*, meaning bell, and the initial function of a clock was to literally chime the hours, indicating, thereby, the times at which the various daily prayers should be read.

Chapter 12

Galileo's Constant Cannonball

It always seems to me extreme rashness on the part of some when they want to make human abilities the measure of what nature can do.

Dialogue Concerning the Two Chief World Systems. Galileo Galilei (1632)

In the modern era it is a high school experiment; in the early 17th century it was an investigation performed at the cutting edge of science. The humble ramp experiment: roll a uniform sphere down a long, angled ramp and measure the time intervals for the sphere to move through various fixed distances. It was a young Galileo Galilei who first began to think about such experiments, although it was an aged and near-death Galileo that finally wrote up the results. Observation and experiment, as far as Galileo was concerned, were the bedrocks of science, and the task of the experimenter was to quantify the relationships that actually exist between physical phenomena. Indeed, Galileo deliberately set out to completely ignore Aristotle and the writings by the other ancient philosophers — the way in which nature worked was not to be found within the theorizing of their words, but in the quantified numbers derived from repeated experimental trials. In his short work entitled *The Assayer*, published in 1623, Galileo forcefully argued that "philosophy is written in this grand book, the universe, which stands continually open to our gaze. But the book cannot be understood unless one first learns to comprehend the language. It is written in the language of mathematics [...] Without which it is humanly impossible to understand a single word of it; without these, one wonders about in a dark labyrinth." Here Galileo is telling us, in no uncertain terms, that we must read the book of nature anew and for ourselves, without recourse to and/or the blind acceptance of ancient doctrine, and while he didn't say it in so many words, from our new readings it is incumbent upon us to deduce the mathematical laws that describe the observed and quantified behaviors.

Galileo's inclined ramp experiment is described in his great work, *Dialogue Concerning Two New Sciences*, published in 1638. It is an artful book, full of new hope for describing the workings of the physical world. And, while the title of his book is overstated — it is not a book about two new sciences; rather, it is a book about a new way of looking at two very old sciences — it is a masterpiece of deduction, argument and literary prose. The *Dialogue* takes place over a four-day time period, and consists of the to and fro banter between three interlocutors: Salviati, Sagredo and Simplicio (of which it is Salviati that carries the voice of Galileo). We are introduced to the ramp experiment on day three of the *Dialogue* in which "change in position" is discussed. Early on in this section, Salviati introduces the mean-speed theorem (without reference to the Merton calculators) and then immediately moves on to his Theorem II, Proposition II, launching into the key result that "the spaces described by a body falling from rest with a uniformly accelerated motion are to each other as the squares of the time intervals employed in traversing these distances." Here, in modern terms, is the embodiment of the statement that the distance s covered by a body in a time interval t while subject to an acceleration a is: $s = \frac{1}{2} at^2$. Later in the day three dialogue Salviati (that is Galileo) introduces us to his actual experimental method, and then generalizes the earlier result. The fundamental point that Galileo observed and makes within the *Dialogue* is that the time for a sphere, no matter what its size, to roll down the ramp is demonstrably independent of its mass. Big or small, for a given ramp inclination to the horizontal (5 degrees, 15 degrees, 45 degrees, and so on) the measured time of motion is always the same. With this experimental result in place, Galileo then takes the experiment to its logical (thought experiment) conclusion and that is, even if the ramp were held vertical, with the sphere now undergoing free-fall motion, it would fall at a rate that is independent of its mass. And this, of course, is the point and conclusion of the fabled Leaning Tower of Pisa experiment — the very experiment we encountered earlier in the Introduction. And here again, of course, is the crucial experimental demonstration that reveals Aristotle and all his subsequent followers as being wrong in their assertion that objects fall at different rates according to their mass, with massive objects falling more rapidly than lighter ones. Indeed, while in the real world a cannon ball dropped from the top of a tower will assuredly fall much faster than a fluffy piece of cotton let go at the same time, the all-important underlying law of nature is that both objects, in the absence of any external forces (such as those due to wind gusts and drag), fall at exactly the same rate.[1] The velocity of free fall, under the action of the Earth's gravitational acceleration g (only later introduced by Newton), will be such that at time t the velocity v will be: $v = gt$, where the acceleration due to gravity is found through experimentation to be $g = 9.81$ m/s^2.

Figure 12.1. Fresco by Giuseppe Bezzuoli (1841) showing Galileo demonstrating his inclined plane experiment to a Medici patron. The theme is purely imaginary, with Galileo (at center) pointing to the apparatus in a classical saint-like pose. The Aristotelian philosophers are shown to the left, consulting their texts, and only the younger audience (center right) is actually looking at the experiment. With consorts at his ear, the Medici patron is locked eye-to-eye with the kneeling priest. Note also the Leaning Tower of Pisa in the background.

---

[1] This result was tested aboard the MICROSCOPE satellite in Earth orbit in 2017, where it was found that two test masses fell, over an 8-day time interval, at exactly the same rate to within two-trillionths of one percent of each other. This incredible result not only confirms Galileo's theoretical reasoning, it additionally reaffirms the great precision of Einstein's theory of general relativity.

Galileo's *Dialogue Concerning Two New Sciences* is a brilliantly reasoned work, and it thoroughly demolishes many long-held Aristotelian perspectives.[2] Not only does Galileo expertly reason that objects in free fall must fall at a rate independent of their mass, he also establishes the law of inertia (more commonly known as Newton's first law [23]) and he correctly describes the parabolic motion of a projectile in terms of two components, one describing the vertical motion and the second the horizontal motion. He also presents a detailed experimental analysis of the pendulum, but incorrectly concludes that the pendulum is an isochronal device (see Chapter 18 below). Conspicuous, however, by its very absence in the *Dialogue* is any discussion of the dynamics of an object dropped into an Earth-crossing tunnel. Given that Galileo had specifically described such a thought experiment in his earlier work *Dialogue Concerning the Two Chief World Systems* (published in 1632), it is a little surprising that the argument is not re-examined. This being said, in the *Dialogue Concerning Two New Sciences*, it is the case that Galileo is specifically questioning Aristotle's views on accelerated motion and employing a reasoned and experiment-based logic to do so. In his *Dialogue Concerning the Two Chief World Systems*, in contrast, Galileo is largely attacking Aristotle's ideas on final causes and comparing the Ptolemaic and Copernican models of the heavens. In this latter context, the Earth-crossing problem has a deep significance since in the Ptolemaic model the center of the Earth coincides with the center of the universe, and according to Aristotle all motion must come to a rest at that location. This was also the conclusion arrived at by Richard Swineshead at Merton College, with, as we have described above, a tunneling object approaching the Earth's core at an ever-slowing rate and never actually arriving at the center in a finite amount of time. In contrast, in the Copernican model, the Earth is the third planet out from the Sun, and having thus lost its central position there is no specific reason why any Earth-crossing object must come to a rest at its center.

It is the same three interlocutors, Salviati, Sagredo and Simplicio, who provide the discussion in Galileo's *Dialogue Concerning the Two Chief World Systems*, and the motion of a body let fall through an Earth-tunnel

---

[2] Which is not to say that Galileo's ideas were instantly accepted and/or adopted by other scholars.

is brought up on three occasions — on days one and two of the dialogue. "Arguing with a certain degree of latitude," Salviati explains on day two, a cannon ball let fall into an Earth-crossing tunnel will have a motion like that of a pendulum, with the velocity gained by the cannon ball in its descent being more than sufficient to carry it right through the Earth's center and that, decreasing in speed, it would thereafter ascend to the far end of the tunnel, repeating the oscillatory motion, up and down the tunnel, for eternity. In this earlier of his *Dialogue*, Galileo works his way to the correct answer by shear thought-experiment reasoning alone, but he can provide no experimental or mathematical argument to prove his case — at the time of his writing, of course, there was no theory of gravitational attraction to aid him in his reasoning. If we adopt the idea that the acceleration acting upon any Earth-crossing object is constant and equal to that which it experiences at Earth's surface, then Galileo, in principle, had at the time of writing his *Dialogue Concerning Two New Sciences* enough information to determine the Earth-crossing time T. Namely, the time t to reach Earth's center, after traveling a distance R, the Earth's radius, will be $R = \frac{1}{2} gt^2$, and the velocity at Earth's center will be $v_{center} = gt = (2Rg)^{\frac{1}{2}}$. The full Earth-crossing time is: $T = 2t = (8R/g)^{\frac{1}{2}}$ which with modern-day numbers yields a time interval of some 38 minutes, and a velocity at the center of $v_{center} = 11.2$ km/s.

Chapter 13

Hooke's Bullet and Newton's Cannon

Nullius in verba

Motto of the Royal Society of London

Robert Hooke provides us with a classic example of the underdog who triumphed. Rising from an impoverished childhood on the Isle of White, by shear grit and genius Hooke worked his way upwards to become a recognized member of England's scientific elite, eventually ascending to the role of Secretary to its most prestigious learned body: The Royal Society of London. Hooke was also the first person actually paid to perform scientific experiments, although the money was not substantial and not always available, and his initial position at the Royal Society was to present to the assembled fellows experiments across the spectrum of the sciences, from biology, to geology, to physics and mechanics. Having been a prominent member since its inception in 1660, Hooke became Secretary to the Royal Society in 1677, and one of his prescribed tasks at that time was to try to entice one of its more taciturn fellows back into the Society's fold. The wayward fellow was Isaac Newton.

To be generous, Newton can be best described as having a difficult personality; he was a genius who felt no compunction to publish his work, and when he did publish he took any criticisms of its content as a personal affront. His approach to writing was along the lines of "here it is, take it or leave it," and, most of all don't bother me with questions about it. Newton's first correspondence with the Royal Society in 1671 concerned his newly designed and constructed reflecting telescope and his thesis *Of Colours*, in which he outlined his famous experiments with glass prisms and described the color properties of light. Hooke made the mistake of criticizing several of Newton's ideas and conclusions concerning the characteristics of light,[1] and taking umbrage Newton withdrew from public debate and further correspondence with the Society. When first contacted by Hooke in November 1679, Newton initially commented that his "affection to philosophy being warn out" he was working in other areas, but since he was being asked to contribute, he continued, "I shall communicate to you a fancy of my own about discovering the Earth's diurnal motion."

The experiment outlined by Newton concerned the experimental verification of the spin of the Earth. Everyone knew, or accepted, at that time that the Earth was spinning on its axis once every 24 hours, but a direct experimental result of this fact was still wanting. Newton proposed that the spin of the Earth could be demonstrated by carefully measuring the displacement that a small "bullet," let fall from a height of "20 to 30 yards," makes with the vertical directly below its release point. Newton argued that the bullet should fall a small distance to the east of the vertical as defined by a plumb line. This was exactly the sort of experiment that Hooke excelled at, and indeed, we read in his diary for January 16th, 1680, "at Garways [*sic*], tryd fall of bullet in the hall with Hunt." This note provides a wonderful early illustration of taking science out into the public arena, for indeed, "Garways" was Garraway's famous London Coffee House, and Hunt was Harry Hunt, Hooke's long-time friend and assistant. Perhaps far from being the ideal place to perform such a subtle experiment (the displacement to be measured was technically of order millimeters), it transpired that Hooke and Hunt felt that they had proved an eastward displacement of the right order. Hooke immediately sent a letter to Newton indicating that [*sic*] "I am now perswaded the Experiment is very certaine, and that it will prove a Demonstration of the Diurnall motion of the Earth as you have happily intimated." It took Newton 11 months to reply to Hooke's enthusiastic announcement, and then his comments were simply that "I am indebted to you thanks." After this

---

[1] These criticisms focused on the exact nature of light — was it composed of corpuscles, as advocated by Newton, or was it some form of wave, as held to by Hooke. This issue was to rage, of course, for several centuries, and it is now one of the founding principles of quantum mechanics in which the wave–particle duality of matter and electromagnetic radiation is a central concept.

letter, there was no more correspondence between Hooke and Newton, and indeed, their relationship only became worse with time. What soured Newton, once again, towards Hooke had been set in place with Hooke's reply to his earlier letter of November 28th, 1679. The offense was that Hooke pointed out that Newton had made a mistake in his reasoning concerning the trajectory of the "bullet" on the basis that after being let fall the bullet continued its motion through the body of the Earth. It was a battle between thought experiments, with the consequence being that Newton secretively set about developing the ideas that eventually resulted, in 1687, in the publication of the *Principia* — perhaps the most famous physics text ever written. In his letter of November 28th, Newton had included a small diagram (Figure 13.1(a)) and described the path of the bullet, arguing that if it could pass through the body of the Earth, it would descend downwards and come to a spiraling stop at the Earth's center. Hooke realized that this was incorrect, and wrote in a letter to Newton, dated December 9th, 1679, that "my theory of circular motion[2] makes me suppose it would be very different and nothing akin at all to a spiral but rather a kind of Elleptueid [ellipse]." It is

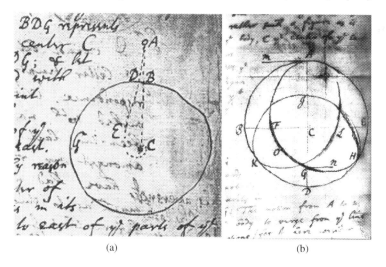

(a)   (b)

Figure 13.1. (a) Newton's original idea for the path of an Earth-crossing bullet — ending at the Earth's center C. (b) Newton's revised path for the Earth-crossing bullet.

---

[2] Hooke's correct idea was that circular motion was a compounded motion, "of a direct motion by the tangent and an attractive motion towards the central body."

clear that Newton realized that he had made a mistake in his earlier reasoning, and replied to Hooke on the December 13th, 1679, with a free-hand sketch (Figure 13.1(b)) showing what is essentially the correct solution — that of a rotating elliptical path (with respect to a fixed observer situated on the Earth's surface). Perhaps all would have gone well between Hooke and Newton at this stage, but Hooke made the further blunder of publishing Newton's initial letter and commenting upon his thought-experiment mistake in describing the continued motion of the bullet. Newton once again withdrew from public debate, but he clearly did not let the subject of compounded planetary motion, as argued for by Hooke, slip from his mind.

Galileo first outlined the idea of compound motion in his 1638 *Dialogue Concerning Two New Sciences*. Specifically, on Day 4 of his discussions, Galileo took up the age-old subject of projectile motion. Galileo explains to his readers that "in the proceeding pages we have discussed the properties of uniform motion and motion naturally accelerated [...] I now propose to set forth those properties which belong to a body whose motion is compounded of two other motions, namely, one uniform and one naturally accelerated [...] this is the kind of motion seen in a moving projectile." Straight away, in Theorem I, Proposition I, Galileo provides us with the key statement that the path of a projectile moving under compound motion will be a parabola. By way of proof, Galileo provides a number of geometrical arguments, and then in a non-quantitative way starts to discuss the effects of air resistance on a projectile. Galileo next sets about describing the characteristic ranges and heights that might be attained by a projectile fired at different angles from the horizontal. In the end, however, Galileo's imagination fails him, and upon discussing the maximum range of any specific projectile he argues that this will never exceed that of a cannon ball, stretching to, perhaps, as far as a few miles. Isaac Newton, however, with a more detailed understanding of dynamics, was soon to pick up upon Galileo's oversight.

One of the problems that limited Galileo in his discussion of projectile motion was the absence of any appropriate mathematics to describe what was going on. With the development of calculus, however, the discussion of projectile motion (in a fixed gravitational field) is relatively straightforward. If the motion of the projectile is composed of a vertical motion in the y-direction and a horizontal motion in the x-direction, then the initial equations of motion are $d^2x/dt^2 = 0$ and $d^2y/dt^2 = -g$, where g is the acceleration due to gravity. These two equations can be integrated to yield the velocity in both the x and y-directions as a function of time t — namely: $dx/dt = V\cos\alpha$ and $dy/dt = V\sin\alpha - gt$, where V is the initial velocity and α is the initial angle that the projectile makes with the horizontal. These two equations can be further integrated to yield: $x = (V\cos\alpha)\,t$ and $y = (V\sin\alpha)t - \frac{1}{2}gt^2$. These latter two equations can be combined to show that the resultant arc of the projectile is a parabola [24], and furthermore they yield expressions for the maximum range and the maximum height of the projectile. Looking at the equation that describes the variation $y(t)$, it is seen that it must be zero at two specific times, firstly when $t = 0$ (corresponding to the launch point) and secondly when it hits the ground at a time t_{fall}, where $V\sin\alpha - \frac{1}{2}gt_{fall} = 0$. The time of flight of the projectile is therefore $t_{fall} = (2V/g)\sin\alpha$. Substituting the flight time into the equation for $x(t)$ then yields the projectiles range $R = (V^2/g)\sin(2\alpha)$, and this has a maximum value when $\alpha = 45°$, giving $R_{max} = V^2/g$. This formula indicates that the larger the initial velocity, the longer the maximum range. The maximum height of the projectile is determined at the top of its parabolic arc where $\dot{y} = 0$, and accordingly it is found that the maximum height is: $H = (V^2/2g)\sin^2\alpha$, with the maximum height being attained at a time $t_H = \frac{1}{2}\,t_{fall}$. When Newton eventually turned his full attention to Hooke's idea, that planets are kept in their orbits by following a compounded motion (as described in Hooke's letter of November 24th, 1679), he was able to intuit, by means of a thought experiment, the connection between the motion of a projectile and the motion of a celestial body.

It was the diplomatic skills of Edmund Halley that eventually coaxed Isaac Newton into producing a mathematical proof that planetary orbits could be explained under the action of a centrally acting force — this force, of course, being gravity. The details were provided by Newton in his *Philosophiæ Naturalis Principia Mathematica*, which first appeared in print on July 5th, 1687, having been paid for, edited and seen through the press by Halley. This work, often described as being one of the most important books to be written in human history, was essentially the result of a meeting between Halley, Hooke and Christopher Wren in January 1684. At this meeting, it is recorded that Hooke claimed that he had mathematically proven that planetary motion was the direct consequence of a centrally acting (that is Sun-centered) inverse square law, and that this same law also allowed for a direct explanation of Kepler's laws [19]. Apparently, neither Halley nor Wren was convinced by Hooke's boast, and accordingly, Halley decided to visit Cambridge in the hope of quizzing Newton on the topic. It was late summer before Halley arrived in Cambridge, and remarkably not only did he find Newton receptive to his questioning, but he also discovered that Newton had already produced such a proof some time before. Newton, however, could not find the details of his calculations so he promised to forward them to Halley at a later date. True to his word, in November 1684, Halley received from Newton a 9-page manuscript entitled *De motu corporum in gyrum* (that is, *Of the motion of bodies in an orbit*). Not only had Newton solved the problem, but he had done so brilliantly and in a wonderfully original manner. Halley immediately contacted Newton with the hope that more material on the topic of dynamics and mathematical philosophy might be available, to which Newton agreed that there was more, much more, and that he would endeavor to write up the details. Over an intensive two-year period of study, Newton turned the brief manuscript that he had presented to Halley in late 1684 into the multi-sectioned, highly detailed, Latin text of the *Principia*. Indeed, Newton went out of his way to make the text difficult, insisting that the reader should toil and struggled with the material, just as he had done so in the writing of it. It is in the *Principia* that Newton presents his three laws of motion, outlines the foundations of classical mechanics, describes his universal law of gravitation and derives, from first principles, Kepler's laws of planetary motion. It is a remarkable book that redefined the way in which natural philosophy was to be conducted, and it indeed determines the foundation moment of modern physical science.

Within the *Principia* and within his earlier *De motu* Newton explained, "that by means of centripetal forces, the planets may be retained in certain orbits, we may easily understand, if we consider the motions of projectiles." It is from this starting point that Newton sets out to describe a thought experiment. If we throw a stone, Newton reasoned, then it will follow (according to his first law of motion) a straight line (or rectilinear) path. The fact that a thrown stone follows a curved path clearly indicates, therefore, that there must be some force continually acting upon the stone to alter its direction of motion — this is the naturally accelerated motion of Galileo and the force of gravity as introduced by Newton. Furthermore, Newton reasoned, if we throw a stone with a greater and greater initial force, then "the farther it goes before it falls to the Earth." "We may therefore suppose," Newton continued, "the velocity to be so increased that it would describe an arc of 1, 2, 5, 10, 100, 1,000 miles before it arrived at the Earth, till at last exceeding the limits of the Earth, it should pass quite by without touching it." In other words, Newton reasoned in his thought experiment that a projectile if propelled with enough force could be turned into a satellite or moon, "and retaining the same velocity it will describe the same curve over and over." Newton famously illustrated his thought experiment with the image reproduced in Figure 13.2.

With his thought experiment and associated diagram, Newton became the first person to suggest that some artificial object might in

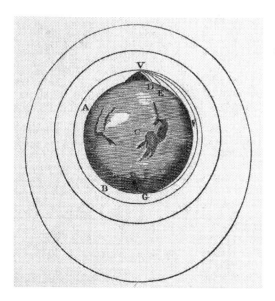

Figure 13.2. Newton's thought-experiment diagram by which a projectile is transformed into an Earth-orbiting satellite. The outer ellipses correspond to objects "fired" from higher mountains (not shown). This thought experiment became reality with the launch of Sputnik 1 on October 4th, 1957.

principle be placed in orbit about the Earth, linking together, as one, the previously separated domains of terrestrial and celestial dynamics. And this, indeed, is one of Newton's key points in the *Principia*: the entire universe is governed by the same physical principles, and natural motion is governed by a *universal* gravitational law. As testament to the legacy of Newton's thinking, in the present day we readily describe universal constants such as the speed of light c, Planck's constant h, the gravitational constant G and the unit element of charge e. Such physical constants define the manner and extent to which the universe can physically work, and if you change any one of the universal constants then you literally change the entire universe. Both modern-day physicists and philosophers of science are deeply concerned with the determination of and the meaning behind the universal constants, asking "Why do they have the values that they do?" [25]

Figure 13.2 offers a very different thought experiment by Newton to that which plagued his correspondence with Hooke in 1679. The projectile now very definitely stops when it hits the Earth's surface (positions D, E, F, and G), but the transition to (in principle at least) everlasting motion is developed through path VBA. It is literally a leap into infinity that is being envisioned. Some 270 years after Newton first expressed the possibility of placing artificial objects into Earth orbit, his thought experiment was realized by the launch of Sputnik 1 on October 4th, 1957.

While Newton did not derive an equation for the speed required to enter into orbital motion, he did develop the mathematical techniques that allow us the means to do so today. This technique is centered on the idea of infinitesimal quantities. Even though the forces driving the motion of some specific object may be variable, Newton argued that if one considers infinitesimal intervals of time Δt then it can be assumed that over that small time interval both the magnitude and the direction of the acceleration can be taken as constants. In this manner any complex dynamical motion can be broken down into a string of many small time intervals and that within each small time interval the laws of constant accelerated motion must apply. Combining this small interval approach with that of additive motions, it is a straightforward matter to determine the minimum speed that an object must have in order to go into orbit about the Earth. Let us imagine a cannonball being discharged in a horizontal direction. With reference to Figure 13.3, in a small time interval Δt

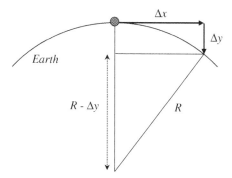

Figure 13.3. The path of an Earth-orbiting cannonball during one brief infinitesimal moment of time Δt. In the time interval Δt, the cannonball travels a distance Δx along a straight-line path, but additionally falls away from the straight-line path by a distance Δy due to the centrally acting force of gravity.

the cannonball will travel a distance of $\Delta x = V\Delta t$, where V is the speed of the cannonball. In the same infinitesimal time interval, because of the Earth's gravitational interaction with the cannonball, it will fall a distance $\Delta y = \frac{1}{2}g(\Delta t)^2$ away from the horizontal. To remain in a circular orbit, however, the distance of the cannonball from the Earth's center must remain constant, and accordingly (with reference to Figure 13.3) an application of the Pythagorean theorem yields the condition that $(R - \Delta y)^2 + (V\Delta t)^2 = R^2$, where R is the radius of the Earth. Expanding the bracket term and cancelling then indicates that $(V\Delta t)^2 = 2R\Delta y - (\Delta y)^2$. The beauty of the infinitesimal idea is that we can safely ignore the $(\Delta y)^2$ term in the equation since it is the square of an already very small quantity, and accordingly it must be even smaller than the original infinitesimal. Substituting now for Δy, we recover the result that the minimum velocity required to place the cannonball into orbit is $V_{orb} = (Rg)^{\frac{1}{2}}$. With this minimum speed, the cannonball will remain indefinitely in orbit, skimming along the Earth's surface (assumed here to be circular). If the initial speed of the cannonball is such $V > V_{orb}$ then its trajectory will follow an elliptical path, with the launch location of the cannon being the perigee (that is, the point closest to Earth) of the orbit. For a circular orbit, the orbital period T_{orb} will be the circumference of the orbit divided by the velocity, giving $T_{orb} = 2\pi(g/R)^{\frac{1}{2}}$, and this equation we have seen before — it is the period of oscillation for an object falling through an Earth-crossing tunnel.

Chapter 14

Newton's Canals

We are certainly not to relinquish the evidence of experiments for the sake of dreams and vain fictions of our own devising; nor are we to recede from the analogy of Nature, which is wont to be simple and always consonant to itself.

Rule III of reasoning in philosophy, *Principia Mathematica*: Issac Newton (1687)

In 1672 while on an expedition to Cayenne, near to the Earth's equator in French Guyana, astronomer Jean Richer noted something strange — his seconds pendulum that had been carefully calibrated in Paris was running slow. As the name suggests a seconds pendulum is one for which the support wire has been carefully adjusted so that its period of swing is 2 seconds, meaning that it takes 1 second to swing from one side of its arc to the other. As we saw earlier, in Chapter 7, the period of a pendulum T is related to the length of its support wire and the local acceleration dues to gravity: $T = 2\pi(l/g)^{\frac{1}{2}}$. In Paris the acceleration due to gravity is determined as 9.809 m/s^2, and the wire length of the seconds pendulum is accordingly 993.86 millimeters. In Cayenne, however, Richer found that his pendulum ran 2 minutes 28 seconds more slowly per day than in Paris. This observation resulted in two important consequences. Firstly, it revealed that the idea of defining the standard meter as being the wire length of a seconds pendulum would not work, since evidently the period of a fixed-length pendulum was not constant everywhere. Additionally, it was a result that caught the attention of Isaac Newton, and he realized that it must be telling us something fundamental about the shape of the Earth.

Indeed, the formula for the period of a pendulum is dependent on two quantities, the wire length *l* and the acceleration due to gravity *g*. If the wire length is held fixed, then any variation in the period must be the result of the pendulum being used in regions where the acceleration of gravity is different. Newton further realized that his equation for the acceleration due to gravity was additionally dependent on two quantities, the Earth's mass *M* and the distance of the pendulum from the center of the Earth *R*, giving $g = GM/R^2$. Taking the Earth's mass to be constant, then the only way in which a fixed-wire-length pendulum's period could change is if the distance of the pendulum from the Earth's center *R* changed from one location to the next. Given Richer's pendulum was running more slowly in Cayenne than in Paris, so the implication was that the acceleration due to gravity in Cayenne was smaller than that in Paris, and this further suggests that Cayenne must be further away from the Earth's center than Paris. In other words, the Earth is not a perfect sphere, but is rather an ellipsoid of rotation, with its equatorial axis being longer than its polar axis. Newton also noted in the *Principia* that "our friend Dr. Halley" also found that the period of his seconds pendulum changed from its calibration location in London and his astronomical observatory, established in 1676, on the island of St. Helena in the south Atlantic Ocean. Newton was convinced that the world was an oblate spheroid, and within the *Principia* he set out to demonstrate just this fact for not just the Earth, but any spinning planet.

That the Earth spins on its polar axis once every 24 hours is an integral part of the Copernican thesis, but in Newton's day there was no unambiguous demonstration of this fact. Certainly, Hooke's performance of the bullet-dropping experiment in 1679 indicated that the Earth seems to be spinning, but it was not a definitive error-free proof. For all this, however, there was no real doubt that the Earth, in spite of our human senses telling us otherwise, must be spinning on its polar axis (as well as moving around the Sun). It was the physical consequences of this rotation that caught Newton's imagination and within the *Principia* he outlined the effects of centrifugal force on the Earth's shape. In Book III, proposition XIX, problem III, Newton jumps right into the debate, writing, "wherefore if APBQ represents the figure of the Earth [Figure 14.1], now no longer spherical, but generated by rotation of an ellipse about its lesser axis PQ; and ACQqca a

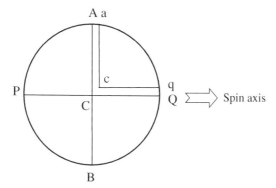

Figure 14.1. Newton's Earth-cutting canal diagram. The Earth's spin axis is to the right in Newton's diagram and the equatorial radius corresponds to AC.

canal full of water, reaching from the pole Qq to the center Cc, and thence rising to the equator Aa; the weight of the water in the leg of the canal ACca will be to the weight of the water in the other leg QCcq as 289 to 288, because the centrifugal force arising from the circular motion sustains and takes of one of the 289 parts of the weight (in one leg), and the weight 288 in the other sustains the rest." Newton doesn't blink for one moment in the development of his grand thought experiment. Rather he simply throws in the idea, without justification and/or a nod to physical reality, that two Earth-crossing canals might be so engineered. For all this, the picture developed by Newton holds the reader's attention and it takes on a sense of an idealized but entirely believable reality.

Newton argues that since the two columns (or canals) of water are imagined to be in contact at Earth's center they must also be in pressure balance. Taking the height of the polar and equatorial canals to be r_p and r_E and assuming that they are both filled with a fluid of constant density ρ, the weight of each column will be: $W_E = \frac{1}{2}\,\rho A g r_E$ and $W_p = \frac{1}{2}\,\rho A g r_p$, the factor of $\frac{1}{2}$ is introduced here to account for the fact that the acceleration due to gravity along each canal varies from a surface value of g to zero at the Earth's center. In the absence of rotation, the Earth would be spherical with $r_E = r_p$ and $W_E = W_p$. With rotation, however, the equatorial canal is in uniform circular motion and it will experience a net inward force due to a centripetal acceleration. In this case, the total forces acting upon the equatorial canal will be $F = W_E - W_P = \frac{1}{2}(\rho A r_E)\omega^2 r_E$, where the $\frac{1}{2}$ has now been introduced to account for the fact that the centripetal acceleration acts upon the fluid's center of mass, which we take to be located at the mid-way point — the bracketed term corresponds to the mass of the equatorial canal and ω is the Earth's angular velocity in radians per second: $\omega = 2\pi/(24 \times 60 \times 60) = 7.27 \times 10^{-5}$ rad/s. Upon substitution for the terms relating to the total force acting upon the equatorial canal, the difference between the polar and equatorial radii is revealed: $(r_E - r_P)/r_E = r_E(\omega^2/g)$. Taking the equatorial radius to be $r_E = 6,378.14$ km, so the difference between the equatorial and polar radii of the Earth is: $r_E - r_P = 22$ kilometers, and this is correct to order of magnitude with respect to modern-day measurements. The flattening of the Earth is given by the expression $f = (r_E - r_P)/r_E = 0.0034 = 1/290$ (Newton derived a value of $f = 1/289$), and likewise, the ratio of the canal masses is of order $W_E/W_P = r_E/r_P = 1.0035 = 287/286$ (this ratio and that for polar flattening are slightly different to that given by Newton purely on the basis of the adopted values for the equatorial radius). When he was writing the *Principia* no one knew what the Earth's flattening might be, and the first expeditions to determine the Earth's shape was not to be made until the mid-1730s (see Chapter 16). Indeed, in 1738 the French mathematician Pierre Louis Maupertuis estimated the flattening to be of order 1/191; just under a century later, however, George Everest used the extensive *Survey General of India* data to determine a flattening of 1/301. The modern value for the Earth's flattening (as adopted by the World Geodetic System: WGS84, and as used in the Global Positioning System (GPS)) is 1/298.257. While Newton did not know what the Earth's actual flattening was, he correctly argued that it should be directly measurable for the planet Jupiter, since it is larger and spins more rapidly than the Earth. Indeed, he estimated a flattening of 3/28, and this value was shown to be of the correct order by telescopic measurements made by Giovanni Domenico Cassini (an avid anti-Newtonian, in fact) at the Paris Observatory in 1691.

The high-resolution satellite mapping that can be performed in the modern era reveals that the Earth is not a perfectly smooth and symmetrical ellipsoid; rather it is an irregular spheroid with multiple humps and bumps. Figure 14.2 specifically shows the Earth's geoid, which is a surface map constructed according to the vertical defined by the

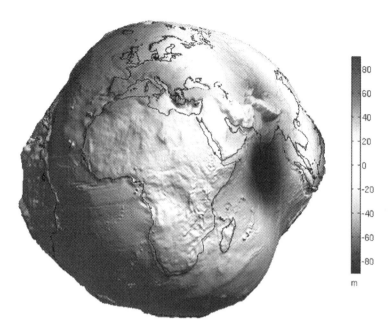

Figure 14.2. The Earth's geoid. This image shows (in a highly exaggerated manner) the surface of an idealized global ocean shaped according to Earth's gravitational field and in the absence of tides. The color scale indicates deviations in height from the idealized reference ellipsoid. Image courtesy of ESA.

direction in which the local gravitational acceleration acts — that is, it is defined by the local, flat surface that a pool of water would adopt. The difference between the geoid and the reference ellipsoid radii, however, are relatively small, amounting to typically less than ±100 meters.

The reality of a smooth sphere, giving way to a smooth ellipsoid, giving way to the bumpy geoid, maps out the corruption of the Platonic ideal form, but the latter geoid is testament to humanity's great mechanical ingenuity and intellectual reach. The geoid illustrates the difference between the perfect ideal as described in the thought experiment and the complex reality of the non-ideal real world. Newton's world was smooth and idealized, stripped of any and all unessential components, but it provides us with a very good approximation to reality, and it reminds us of the fact that while the Devil is often in the details, it is not always the small details that are fundamental to the development of understanding.

Chapter 15

Halley's Hollow Earth

You are to make the best of your way to the Southward of the Equator, and there to observe on the East Coast of South America, and the West coast of Affrica [sic], the variations of the Compasse [sic], with all the accuracy you can.

Admiralty instructions to E. Halley, Captain of HMS Paramour.

Edmund Halley was a man of action. Scholar, sea captain, mathematician, astronomer, international spy and diplomat, Halley lived an extraordinary life. While he tends to be best remembered in the modern era for his triumphant prediction of the return of the comet named in his honor, this is perhaps one of his lesser (great) accomplishments. A number of portraits of Halley exist, but it is one produced by the Swedish-born artist Michael Dahl circa 1736 that is of particular interest here. Dahl's portrait is rather formal, and it depicts a somewhat stern-looking octogenarian Halley dressed in formal robes. In his right hand Halley holds a single sheet of paper, upon which is shown a diagram consisting of a set of three rings centered upon an inner sphere (Figure 15.1). It seems an odd kind of device to be included in a portrait, but it reflects one of Halley's most interesting and audacious ideas — indeed, an idea that he had put into print some 44 years earlier. The diagram in Halley's hand is that of a hollow Earth, or to be more precise a model of the Earth containing two inner shells, with diameters corresponding to Venus and Mars, surrounding a central core having the same size as Mercury. It is a highly ordered model, containing no fire caverns and lava tubes of the kind imagined by Kirchner (recall Figure 4.1), and, of course, it is entirely wrong. For all this,

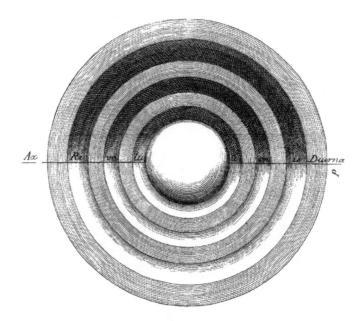

Figure 15.1. The hollow-Earth model of Edmund Halley. The Earth's spin axis is shown horizontally, and the interior contains two shells and one central sphere.

however, it was a brilliant attempt to describe the Earth's magnetic field, and especially its measured variations over time.

That the Earth behaved like a giant spherical magnet was first demonstrated in one of the strangest books ever written: *De Magnete* by William Gilbert. Published in 1601, Gilbert's book was all together something different. Part hard-core science, with experiments being outlined, conducted and then explained in minute detail, and part shear fantasy, with the conclusion being drawn that the Earth was fully alive and that the geomagnetic field was the physical embodiment of the Earth's soul. Halley held no truck with Gilbert's live Earth, but he did have a deep, lifelong fascination with the Earth's magnetic field. As Captain of HMS Paramour, Halley had conducted a whole series of magnetic measurements[1] at different locations across the north and south Atlantic Ocean

---

[1] Technically Halley was interested in what is called magnetic variation, or magnetic declination, which is defined as the angle between the directions of magnetic north and true geographic north. This angle continuously varies according to location on the Earth and over time.

during the late 1690s, and he carefully collected similar such magnetic measurements from other observers and sea captains from around the world. Halley's seafaring and meticulous study resulted in the publication, in 1701, of his *General Chart of the Variation of the Compass*, a chart that revealed for the very first time isogonic lines of constant magnetic declination. But now we are a little ahead of ourselves. Halley's *General Chart* was the product of numerous direct measurements and hard interpretive graft, but the underlying model that Halley proposed, the one shown in his portrait by Dahl, to explain the observed magnetic variation was something altogether daring and imaginative, and first described in a 1692 paper published in the *Philosophical Transactions of the Royal Society of London*. The paper has the rather lengthy, but entirely descriptive title of *An account of the cause of the variation of the magnetic needle with an [sic] hypothesis of the structure of the internal parts of the Earth*. In this paper Halley writes, "We have adventured to make the Earth hollow and to place another globe within." The idea is fantastic, but it is a model based upon sound scientific reasoning and observational facts — the fact that it is neither the correct, nor the present-day model for describing magnetic variation is not the point here. Rather, it is an extreme result predicated upon the available knowledge of the time. Halley's hollow-Earth model, at its heart, is not a gimmick or fantasy-based concept, but it is a serious attempt to account for observations, puzzling observations at that, relating to the Earth's magnetic field.

The results of Halley's first foray into the deciphering of the Earth's magnetic field were presented to the Royal Society in 1683. He gathered together information about the magnetic variation from as many locations as he could from all around the world, and after sifting through the data he wrote, "I can come to no other conclusion than that, the whole globe is one giant magnet having 4 magnetic poles." That there were apparently 4 magnetic poles is entirely at odds with the ideas presented earlier by Gilbert, and those deduced experimentally from *Terrella* (small spheres ground from naturally magnetic loadstone). The result also indicates the complexity that is inherent in the structure of the geomagnetic field. Halley argued that there were two northern poles, one to the north of Banks Island in the Beaufort Sea, and one to the north and west of Greenland in the Arctic Ocean. The two southern poles were located in Antarctica, one in the region that is now called Ellsworth Land and the other close to the location of the present-day Vostok Station in Wilkes Land. Not only did Halley conclude that the Earth's magnetic field was multipolar, he also presented evidence to indicate that the magnetic poles were undergoing a slow westward drift. In these early stages of his investigations Halley present no specific model to explain what might be going on; rather arguing that the reasons for the variation were "secrets as yet utterly unknown to mankind; and are reserved for the industry of the future ages." It was Halley himself that eventually provided the industry to take the observations to the next, theoretical modeling, level. A number of important events had to happen first, however. Firstly Halley had to undertake his voyages aboard the *Paramour*, and second Newton, with Halley's prodding and financial assistance, had to produce his *Principia*.

With reference to his earlier 1683 magnetic variation results, Halley introduced his 1692 review with the comments that he had almost despaired "of ever being able to account for this phenomenon." For all this "despair," however, Halley soon outlined his idea that the four observed magnetic poles are produced in two different regions of the Earth. The one set is anchored to the surface of the Earth, viewed as a hollow shell, and the second set is anchored to an inner sphere (or "magnetic nucleus") that slowly rotates with respect to the stationary, outer shell, magnetic field. The model, in the mind's eye, can be made to explain the observations and, if appropriately parameterized, could be constructed to account for the observed polar drifts (although Halley did not try to enumerate this last step). Having split the Earth into an outer shell and inner nucleus, Halley then sets about exploring the consequences of the model. He worries about what use such an inner structure might have in the divine sense of providing for life. He argues, for example, that perhaps there are luminescent rocks under the outer shell to provide light to the inner shell beings (whatever they might be), and that the space between the inner core and the shell might provide for a sustaining atmosphere. He also argues that a particularly odd and unexpected result derived within Newton's *Principia* might also be explained by the hollow-Earth idea. Specifically, within the *Principia*, Newton had developed a tidal theory based upon the gravitational pull of the Moon and the Sun, and he had used this theory to estimate the density of the

Moon. Strangely, Newton found that the density of the Moon was nearly twice that of the Earth, "or as 9 to 5 nearly." This was odd since the Moon was (reasonably) assumed to be made of the same material as the Earth, and so accordingly should have a similar deduced density. Newton's result was in fact in error (one of his rare blunders), but Halley saw that by making the Earth hollow one could then reconcile the apparent density difference. Taking the Moon to be solid through and through and made of "earth, water and stone,", the apparent low density of the Earth could be explained by making it hollow — this effectively increases the volume of the Earth and thereby reduces (for a fixed mass) its density. Having argued that the Earth is hollow, but that the Moon is solid, Halley then effectively invokes a uniformity clause — or, as Halley put it, "I have adventured to add the following scheme." If the Earth is hollow, then why not let all the planets be hollow, and hollowed-out in a uniform fashion. Accordingly, although not specifically needed to explain the magnetic variations, Halley argues that inside of the hollow shell of the Earth (estimated to be 500 miles in thickness) is another hollow sphere the size of Venus, inside of which is another hollow sphere the size of Mars, and finally right at the very center of them all is a solid nucleus the size of Mercury. Although not articulated, we suppose Halley envisioned that the other terrestrial planets were equally sub-divided: Mars having just an outer surface shell and an inner Mercury-sized core; the outer shell of Venus containing an inner Mars-sized shell, and central core the size of Mercury.

Halley's hollow-Earth model is new, speculative and highly exciting. It offers possibilities for further development; it offers possibilities for further understanding the structure of the planets within the solar system; and it explains why the Moon has an apparent density nearly twice that of the Earth. Halley's theory does exactly what a scientific theory should do, it explains the observed facts, it explains the observed polar drift, it in principle provides predictions concerning future variations in the geomagnetic field, and it provides for future refinement. With all this going for it, it is perhaps a little surprising that, as scientific theories go, it was a complete flop — nobody wanted to work with it and/or develop its predictive potential. Perhaps it is because of this rejection that Halley looks so glum in his portrait of 1736.

That the Earth has a magnetic field was a discovery of profound importance and from a modern perspective its very existence provides us with important clues as to the nature of Earth's deep interior. Who first noticed the seemingly magical properties of the suspended loadstone to orientate itself in the same direction is not known. It was certainly a phenomenon known to ancient Chinese navigators, but its connection with anything relating to the Earth itself is all Gilbert. Why the Earth should have a magnetic field, apart from Gilbert's suggestion as an emanation of a benevolent soul, is another story. Indeed, there is something altogether strange and spooky about magnetism, and it still remains unnerving, at least to the author, to feel the invisible forces and tensions that *materialize* when two magnets are brought together. For more than 200 years after Gilbert's publication, the otherworldly quality of magnetism continued to hover unexplained and inexplicable. That magnetism was related to another phenomenon, electricity, was only realized, partly by inevitable accident, by Danish physicist Hans Christian Oersted in 1820.

On April 21st of that fateful year, Oersted had agreed to perform a set of demonstrations to some friends concerning the heating of a wire through which an electric current was being passed. He noticed, however, that every time that he connected and disconnected the switch to the battery, the needle of a compass that chanced to be on the table moved. Oersted had discovered the fact that a moving electric current produces a magnetic field. Within a year Michael Faraday, at the Royal Institution in London, had experimentally demonstrated the reverse Oersted effect by showing that a magnetic field could deflect an electric arc. Meanwhile, in France, Andre-Marie Ampere set out, with great zeal, to describe all magnetic phenomena in terms of tiny loops of circulating electric currents. The real theoretical breakthrough, however, was presented by James Clerk Maxwell, when in 1865 he published his landmark paper, *A Dynamical Theory of Electromagnetic Fields*. Read to the assembled fellows of the Royal Society of London on December 8th, 1864, the published paper changed theoretical physics for all time. The work was highly mathematical, and it invoked the use of field equations and it was utterly brilliant. Not only did Maxwell's equations successfully conjoin electricity and magnetism, he also found that electromagnetic waves must move through space at the speed of light, and this indicated that light itself must be a form of electromagnetic radiation. For all these incredible realizations, however, the mechanism responsible for the origin of Earth's magnetic field remained a mystery.

Ampere had suggested that electric currents moving within the Earth's interior might be the source of its magnetic field, but there is more than one problem with this idea. One such problem follows from the studies by German mathematician Carl Friedrich Gauss, who reasoned that the Earth's magnetic field could not be due to some permanent magnetic mass within its interior. This conclusion was based upon the fact that not only did the field strength vary, but also, as Halley had mapped out, its orientation with respect to the Earth's surface changed with time. Furthermore, even permanent magnets are not necessarily permanently magnetic, and it can be shown that the characteristic decay time for the Earth's magnetic field to entirely dissipate is of order 20,000 years. Accordingly, we either live in a very special epoch when the Earth just happens to have a magnetic field, or, as is indeed the case, the Earth's magnetic field is continuously regenerated.

That Earth's magnetic field might be the result of a self-sustaining dynamo mechanism was first outlined in theoretical detail by Irish physicist Joseph Lamor in 1919, although it was to be the mid-1940s before Walter Elasser in the United States was able to work out most of the geophysical details. Indeed, Elasser developed the idea that the Earth's dynamo was powered by convection currents in the liquid outer core (located at depths between 2,900 and 5,100 km below the Earth's surface). That the Earth even had a liquid metallic outer core was only realized in the early 1900s, and its existence was first hinted at by Irish physicist Robert Dixon Oldham on the basis of seismological recordings of distant, energetic earthquakes. It would be a great overstatement to suggest that modern-day geophysicist fully understand the many processes at play in the generation of the Earth's magnetic field, but a highly plausible picture has been built-up (Figure 15.2). The key ingredients are heat, convection and rotation. Currents of conducting liquid metal are developed within the Earth's outer core and the Coriolis force (due to the spinning Earth) organizes these currents into giant columns of circulating electric charge. These circulating currents in turn generate a magnetic field, and provided that the Earth's core remains hot the self-sustaining dynamo will keep operating.

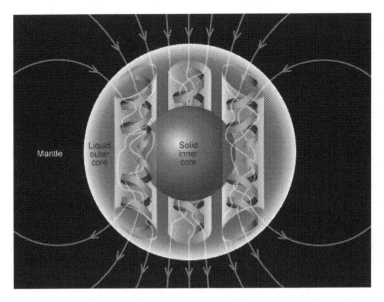

Figure 15.2. A schematic diagram of Earth's magnetic dynamo. A combination of heat flow from the center, convection within the outer core and rotation enables a self-sustaining dynamo mechanism to operate. Image courtesy of the USGS.

While the Earth's magnetic field appears to be described by a self-sustaining dynamo mechanism, this does not mean that its behavior is predictable over time. Indeed, the record that is frozen within Earth's more recently formed magmatic rocks indicates that the magnetic field has undergone dramatic reversals on many occasions during the past many hundreds of millions of years. The last reversal, which lasted an unusually short 450 years, occurred some 41,000 years ago, when the north and south magnetic poles switched orientation. The switching cycle, however, appears to be largely unpredictable with the period varying from as short as 100,000 years to as long as 1 million — and no one knows for sure when the next reversal might take place.

The numerous reversals displayed in Earth's magnetic field do not appear to have had any long-term or wide-spread affects upon animal life, even though it is now known that many migratory animals use the Earth's magnetic field to orientate themselves. There are no clear and/or agreed upon mass-extinction events observed within the fossil record that correlate with the magnetic field reversals, but it is reasoned that life on Earth is crucially dependent upon the Earth having a magnetic field. The geomagnetic field protects the atmosphere from the solar wind and the particulate strafing provided by cosmic rays. In this latter respect the panned Hollywood movie *The Core* (recall Chapter 9) was at least correct, and the existence of a magnetic field about a planet is now considered to be a basic requirement, from an astrobiological standpoint, for there being any hope of life evolving somewhere else, beyond the Earth, in the galaxy. For all this, however, the Earth's inner dynamo and the flip-flop polarity of the geomagnetic field may be responsible for our very existence.

This controversial possibility has been developed by Joseph Meert (University of Florida in Gainesville) and co-workers who have noted a magnetic oddity at the time of the Cambrian explosion, dated to some 542 million years ago, when life, as portrayed in the fossil record, truly took off in ever more inventive and convoluted evolutionary directions. Indeed, this is the time and geological strata to which one of our most distant known ancestors, *Pikaia gracilens*, is to be found (Figure 15.3).

Figure 15.3. *Pikaia gracilens*, one of the many new forms of life found in the extensive fossil beds of the Burgess shale of British Columbia. *Pikaia* measured just a few centimeters in length and is believed to be one of the earliest ancestors to all vertebrates. Image courtesy of Wikimedia Commons.

Not only does the fossil record within the Burgess shale tells us that something truly remarkable was taking place with respect to evolutionary exploration, but the geological record further indicates that the Earth's dynamo was in a hyperactive polarity switching mode. Meert and co-workers argue that the paleomagnetic indicators reveal that both the Earth's magnetic field strength was low at the onset time of the Cambrian explosion and that the polarity reversal rate was at least 20 times higher than at present. This low field strength and rapid polarity switching, Meert suggests may have resulted in the destabilization and destruction of the ozone layer,[2] thereby allowing a higher-than-normal solar UV flux to penetrate to the Earth's surface. Since solar UV radiation is known to cause DNA damage and mutation, it has been suggested that any adaptation allowing for mobility and/or the growth of hard surrounding shells would have been highly advantageous and accordingly strongly selected for in an evolutionary sense. *Pikaia*, with its rudimentary notochord, evolved and swam within the Cambrian era ocean, moving like a smaller version of the modern-day lamprey, and it is from this small worm-like wonder that the chordate phylum, to which we *Homo sapiens* belong, eventually evolved.

---

[2] The detection of ozone (O_3) within the atmosphere of a distant exoplanet would be a clear indicator, in fact, of life having evolved there.

Chapter 16

Dr. Akakia's Diatribe and Euler's Miracle

A true philosopher does not engage in vain disputes about the nature of motion; rather, he wishes to know the laws by which it is distributed, conserved or destroyed, knowing that such laws are the basis for all natural philosophy.

Les Loix du Mouvement et du Repos, déduites d'un Principe Métaphysique.

P. L. Maupertuis (1746).

Pierre Louis Maupertuis was admitted to the prestigious French Académie des Sciences in 1723. By this time, at the age of 25 years, he had established a name for himself as a gifted mathematician and a strong continental advocate for the new ways of conducting science based upon the philosophy expressed by Newton. For all this, however, Maupertuis had a reputation for being difficult and argumentative — a predisposition that won him several distinguished rivals. One of his first inter-personal disputes, however, resulted in an important series of expeditions and measurements that would determine the shape of the Earth and vindicate Newton's brilliant perceptions concerning gravity. This particular dispute had its genesis in the early 1730s and it involved Maupertuis in a disagreement with Jacques Cassini, a fellow member of the Académie. At issue was the shape of the Earth — was it a perfect sphere, was it an oblate sphere (as predicted by Newton) or was it a prolate sphere (as argued for by Cassini)?

Newton had demonstrated in his *Principia* that the Earth must bulge at the equator (recall Chapter 14), making it an oblate sphere, while

Cassini argued that the Earth was extended along its spin axis, making it a prolate sphere. The argument was predicated on just how to interpret a series of pendulum measurements made at different latitudes on the Earth's surface. It was seen earlier that the period of a pendulum is determined by its length l and the local gravitation attraction g, and accordingly any variation in a pendulum's period of oscillation must be due to changes in the length of its support wire or in the local gravity which is determined by mass and radius of the Earth: $g = GM/R^2$. If the Earth was a perfect sphere with a uniform distribution of matter within its interior then g would be constant at all and every location on its surface. Extensive sets of measurements with fixed-length pendulums, however, had repeatedly shown that they ran more slowly in the equatorial regions — that is they completed fewer swings per day than a similar-length pendulum located close to the Polar Regions. This observation implies that the gravitational acceleration must vary over the Earth's surface, and be somewhat smaller in the equatorial regions. Given Newton's formulation, a smaller gravitational acceleration g can be realized by either reducing the mass M term or increasing the radius R term. The argument between Maupertuis and Cassini centered upon which term to vary — Maupertuis followed Newton and argued that the gravitational attraction was smaller at the equator, because the Earth's equator bulged outwards (giving a larger R term), thus making it oblate. Cassini, in contrast, argued that the gravitational attraction was smaller in the equatorial regions because of a reduction in the mass term, and he accordingly asserted that the Earth must be prolate. Of our two august *savants*, the question is, of course, who had the right explanation? The answer was eventually found through geometry measured large on the landscape — that, and Royal patronage. The French Geodesic Mission was founded under the guidance of the Académie des Sciences and the funding of King Louis XV, in 1735, and Maupertuis was made its director. Two expeditions were organized. One group was to sail for Ecuador and make detailed geodetic measurements to determine the length in kilometers of one degree of arc at the Earth's equator. The other group was to sail for Lapland, close to the arctic circle, and make similar such measurements for the length of arc at that location. If the Earth was oblate then the equatorial arc length should be longer than the length of the Lapland arc. Conversely, if the length of the arc at the equator was smaller than that at Lapland then the Earth must be prolate. These were heroic expeditions, fraught with many dangers but inspired by the idea of scientific discovery.

The expedition to Ecuador was almost a comedy of errors — the scientific personnel argued amongst themselves, split up and regrouped, and then argued some more, and what was supposed to be an expedition lasting a year or two turned into a nine-year (and more for some) odyssey. The Lapland expedition, which set out in 1736 under the direction of Maupertuis, was a relatively quick affair, and within two years their results were in. The length of the northern one-degree meridinal arc was shorter than a similar arc measured in France (and eventually it was shown to be shorter than the measured arc length at the equator). The Earth is an oblate sphere, just as Newton and Maupertuis had argued for. While Maupertuis is rightly lauded for his work *Figure of the Earth determined by observations ordered by the French King at the Polar Circle* (published in 1738), he is additionally remembered to this day for his fundamental ideas relating to the principle of least action. This latter concept, however, saw Maupertuis involved in yet more public controversy, and in this case he was not to triumph over his rivals.

The principle of least action is essentially a statement about the way in which the universe works. That is, it acts towards the minimization of relevant quantities. Maupertuis introduced a quantity called *action*, which was expressed as the product of the mass, distance traveled and the velocity. With this *action* Maupertuis argued that the dynamics of a single particle, or indeed, that of a whole assembly of particles, could be described. He further argued that the behavior of light, as it traveled through different optical media, could be explained according to the adoption of the shortest path length.[1] Maupertuis published several key papers on his least action principle between 1741 and 1746, and at this same time he received several prestigious honors and appointments. Indeed, in 1742 Maurpertuis was made Director of the Academy of Sciences, and in 1746 he was invited by Frederick II of Prussia to become

---

[1] In this way he was able to explain Snell's law of refraction. Another French mathematician, Pierre de Fermat, had previously cast Maupertuis's action in terms of light taking the shortest travel time.

Figure 16.1. Engraving of Pierre Louis Maupertius by Jean Daulle (1741). Note the reindeer and sled towards the bottom of the engraving (representing the Lapland expedition), and that Maupertius, in order to illustrate that it is an oblate spheroid, is symbolically squashing the Earth at the North Pole with his right hand.

President of the Royal Prussian Academy of Sciences. From Berlin Maupertuis published yet more papers on the principle of least action, and it was in Berlin that the seeds of his demise were sown.

The story and the inter-personal machinations are complex, but it transpires that in 1751 Johann König (who Maupertuis had earlier proposed for election to the Berlin Academy) published a paper (written much earlier, in fact) that suggested Maupertuis had plagiarized his principle of least action from the writings of Gottfried Leibniz. Maupertuis was not well pleased with the charge of plagiarism,[2] and used his powers as President of the Prussian Academy to expel König from all scientific society and discourse. At this juncture Voltaire enters the narrative. Maupertuis and Voltaire had known each other for some considerable time, but Voltaire saw the drive by Maupertuis to ostracize König from the fold of the Academy as the suppression of youthful talent by a bigoted and dictatorial high official. Thus riled, Voltaire set about seeking Maupertuis's downfall. In this revenge, Voltaire was helped in his endeavors by the publication of a small pamphlet by Maupertuis entitled *Lettre sur le progress des sciences* in 1752. Maupertuis's *Lettre* is a wonderfully crafted essay that outlines possible directions for future research programs and mega-projects that the King might wish to provide patronage and financial support. Amongst the topics that Maupertuis set out to promote are the development of new telescopes to measure stellar parallax, the exploration of Africa and Australia, the search for the North West passage, the study of medicine and animal classification, and the construction of a tunnel to the center of the Earth. The latter, of course, would be a mega-engineering project, but it would provide both valuable mineral extracts as well as valuable scientific knowledge about the structure of the Earth. Voltaire seized upon the publication of the *Lettre*, and with all his wit and satirical skill he set about lampooning Maupurtuis in a short pamphlet entitled *Diatribe du docteur Akakia* (published in 1752). Voltaire's pen was full of venom, and with respect to the Earth-tunnel project he writes, "we must further inform him [Dr. Akakia, i.e., Maupertuis] that it will be extremely difficult to make, as he proposes, a hole that shall reach to the center of the Earth (where he probably means to conceal himself from the disgrace to which the publication of such absurd principles has exposed him). This hole could not be made without digging up about three or four hundred leagues of Earth; a circumference that might disorder the balance of Europe" [26]. The *Lettre* of Maupertuis is full of hope and is a broad-minded vision concerning scientific enquiry, whereas Voltaire's *Diatribe* is full of sarcasm and blinkered opinion, but, as is typical in human history, Voltaire's work won the day. As a broken

---

[2] It was shown long after the death of both Maupertuis and König, that the charge of plagiarism was entirely unfounded and that Maupertuis developed his ideas independently to those of Leibniz.

and embittered man, Maupertuis resigned his Presidency of the Prussian Academy in 1757, and died just a few short years later.

While Maupertuis was the main target of Voltaire's *Diatribe*, the great Swiss mathematician Leonhard Euler was also caught up in the controversy, although Euler managed to survive, his reputation intact, most of the backlash. Euler had taken up his post at the Berlin Academy in 1741 and for the next 25 years set about developing what have become foundational works in calculus, number theory[3] and geometry (Figure 16.2). Euler was certainly on friendly terms with Maupertuis during the latter's Presidency at the Academy, and while it is not clear where and when Maupertuis developed the idea to cut a tunnel to the Earth's core, as outlined in his *Lettre*, he may well have obtained the essential idea from Euler.

Euler's relationship with the Earth-tunneling problem began before his move to the Berlin Academy. Indeed, as a young and aspiring mathematician, at the age of 20 years, Euler added the analysis of motion within an Earth tunnel to his habilitation thesis presented to the University of Basel in 1727. A habilitation thesis was an early form of *curriculum vitae*, and it was submitted as part of the application process for

Figure 16.2. German commemorative stamp showing the profile of Euler and recording his famed polyhedral formula: $e - k + f = 2$, where e is the number of vertices, k is the number of edges and f is the number of faces.

---

[3] Readers of the *Mathematical Intelligencer* journal voted Euler's remarkable relationship $\exp(i\pi) + 1 = 0$ to be "the most beautiful mathematical formula ever." Beauty, of course, is all in the eye of the beholder, but it is without doubt the case that Euler's formula is wonderfully intriguing, containing, as it does, two irrational numbers (e and π), two integers (0 and 1), and an imaginary number (i being the square root of minus one).

admittance to an academic post. The thesis was intended as a show piece, an example of what the candidate could do, and its subject matter was used as the basis for a public examination. The partial title of the thesis that Euler submitted for consideration runs as follows: "May it bring you happiness and good fortune — Physical dissertation on sound which Leonhard Euler, Master of the liberal arts submits to the public examination of the learned [...] at the free professorship of physics by order of the magnificent and wisest class of philosophers whereby the divine will is nodding ascent [...]" To this main thesis, which essentially considered the propagation of sound in air, Euler attached a set of additional notes, one of which considers the thought experiment of dropping a stone into a tunnel cut through the Earth. Euler writes [27], "from the position of the center of Earth (which moreover is truly far away and alien) any bodies are to be attracted by the inverse square of their distance, and a hole is to be drilled through the center of the Earth. It is asked, for a stone sent down the hole, what will happen, when it reaches the center, whether it will either remain there permanently, or progress away from the center without pausing and return to us soon from the center of the Earth. I confirm the latter is the case." At this stage it might seem that Euler had in mind something like Galileo's oscillating cannonball solution to the Earth-tunneling problem, but a closer examination of Euler's writings reveals an altogether different outlook. Indeed, without discussion and/or proof Euler argues that three solutions to the Earth-tunneling problem are possible: (i) the stone will come to a rest at the Earth's center [this is the Aristotelian and Merton calculators' expectation]; (ii) the stone will proceed beyond the center [this is Galileo's expectation]; and (iii) the stone, having acquire an infinite velocity at the Earth's center, will instantaneously return from the center to the surface point from which it had been dropped. Of all the answers that one might expect to be given, Euler indicated that solution (iii) would apply, and while this might seem strange to contemplate it was, in fact, a reasoned result. Much to Euler's great disgust, however, it transpired that he was not offered the physics professorship at Basel.

We learn more about Euler's reasoning concerning the Earth-tunneling problem in his *Mechanica sive motus scientia analytice exposita*, published in 1736. In this remarkable two-volume work Euler describes the mathematics relating to particle motion, and within its pages he specifically describes the motion of a body moving under the action of a central force, $F(r) = \alpha r^n$, that varies as the distance r from the center raised to some power n, where n can be any number, positive or negative, and where α is a constant. The Earth-tunneling problem is described by a one-dimensional equation (derived according to Newton's second law — see the Appendix) in which $m(d^2r/dt^2) = F(r)$, where m is the mass of the object. This equation indicates that the rate of change in the velocity (dr/dt), that is, the acceleration of the body of mass m at location r is directly proportional to the applied force at r. Both special and general solutions to this equation of motion can be found. If, for example, $n = 0$, then $F(r) = \alpha$, and we have the situation of a body moving under a constant acceleration — this is the condition that applies in Galileo's solution to the Earth-crossing problem. When $n = 1$ and $\alpha < 0$, we recover the equation for simple harmonic motion (see C.P. No. 1's solution in the Appendix). For all $n \geq 0$, the force goes to zero at the center where $r = 0$, but for all $n < 0$, the force becomes infinite at $r = 0$. It is this latter singularity condition that forced Euler to consider the possibility of infinities arising under Newtonian dynamics, since the gravitational force varies as $n = -2$. So, by Euler's reasoning, an object let fall into an Earth tunnel would experience a greater and greater acceleration as it moved closer and closer to the Earth's core (the $r = 0$ location), acquiring in consequence a higher and higher velocity, until at the very center the body would be moving infinitely fast.[4] Sensing a distinct disconnect between the mathematical result and reality, Euler writes of this latter condition, "this infinite step is to be deplored." Interestingly, Euler's apparent infinity is in the exact opposite sense to the infinity derived by the Merton calculators, who deduced that the body would slow down as it approached the center of the Earth, only reaching the center after the passage of an infinite amount of time. The question now, however, as far as Euler was concerned, is what happens to the body once it has acquired an infinite speed. Indeed, he argues that it can go anywhere

---

[4] This work was performed long before Einstein introduced the concept of the limiting speed of light in his 1905 publication on special relativity. Indeed, Euler was writing 40 years before the finite speed of light was first demonstrated, via observations of Jupiter's moons, by Danish astronomer Ole Rømer.

instantaneously, and therefore, why not let it jump back instantaneously to its starting place — which was answer number (iii) in his habilitation theses. Euler further comments in his *Mechanica* that "here the calculation rather than our judgment being trusted and established, the jump if it is made from the infinite to the finite, is not thoroughly understood."

Through Euler's analysis in the *Mechanica* we essentially reach a philosophical impasse: do we accept the strict mathematical analysis and its strange physical implications, or do we insist that physical infinities are impossible and therefore the mathematical analysis must be wrong, or at the very least incomplete. Euler never wavered in his viewpoint that the mathematics was right, and that some form of singularity and infinite velocity must be accounted for. In actuality, however, while Euler did provide a correct set of mathematical solutions to the problem that he outlined, the problem, as such, is not anchored in reality. Euler's essential mistake was to take the α term in the force law as being a constant, thereby enabling a singularity to occur at $r = 0$. In the real world situation, when the effect of Newton's shell theory is applied and when the density is necessarily finite at the center, the α term approaches zero more rapidly than the radius squared term r^{-2} approaches infinity, with the net result that at the center of the Earth, the gravitational force will be zero, $F(r = 0) = 0$, rather than infinity, $F(r = 0) = \infty$. It is generally taken as being a pathological failure when a theory predicts the attainment of infinities, since such results are clearly contradicted by reality (where infinities do not occur). This being said, there are additionally times when the infinities predicted within a theory can be *diffused* mathematically and *transformed* thereby to yield finite results — such is the case, for example, in the use of renormalization techniques in quantum field theory.

Chapter 17

Collignon's Slant

Slice him where you like, a hellhound is always a hellhound.

P. G. Wodehouse

To this point we have essentially assumed that any Earth-crossing tunnel will be diametrical and automatically pass through the Earth's core. This condition, of course, need not always be the case — shorter tunnels through the interior of a sphere can readily be envisioned, and the first person to consider the non-center-passing Earth-tunnel problem was French engineer Édouard Charles Romain Collignon (1831–1913) [28]. Unlike earlier situations where the Earth-tunnel problem was essentially viewed as a theoretical, thought-experiment problem, Collignon was interested in developing a real-world, mass-transit system — indeed, a system and a vehicle that has now come to be known as the gravity train. Collignon was a railway engineer[1] and he played an important role in the construction of the rail links between St. Petersburg and Warsaw, and Moscow to Nizhny Novgorod. In 1872 he was one of the founding members of the Association Française pour L'avancement des Sciences, indeed, acting as its President in 1892. At the 1882 meeting of the Association held in La Rochelle, however, Collignon presented (on the 28th of August) his idea for a tunnel system that could transport passengers across the entire Earth in just 42 minutes. Collignon began his paper (Figure 17.1), entitled simply as *Problème de Méchanique*, by considering

---

[1] This being said, he is also remembered for the equal-area pseudocylindrical map projection — the Collignon projection — and he wrote several texts on analytic mechanics and the Russian railway system.

M. Ed. COLLIGNON

Ingénieur en chef des ponts et Chaussées.

PROBLÈME DE MÉCANIQUE

— Séance du 28 août 1882. —

§ 1ᵉʳ

Le problème que nous allons traiter a pour premier objet le *mouvement d'un point pesant sur une corde de la sphère terrestre*. Nous supposerons que la densité du globe est partout la même ; la pesanteur varie alors à l'intérieur proportionnellement à la distance au centre. Nous supposerons en outre que le glissement du point développe un frottement proportionnel à la pression normale qu'il exerce sur la droite directrice, et que le coefficient f de ce frottement est connu.

Soit   O le centre de la sphère terrestre ;

         OA $=$ R, le rayon du globe ;

         AB, la corde suivant laquelle le mouvement doit avoir lieu ;

         α, l'angle BAA' que cette corde fait avec le diamètre AA' de la sphère, mené par le point de départ A du mobile.

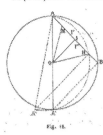

Fig. 11.

Le point pesant est abandonné sans vitesse au point A. Il glisse dans la direction AB. Au bout d'un certain temps t, il se trouve en M, à une distance $\mathrm{IM} = x$ du milieu I de la corde AB. La pesanteur agit sur le point dans la direction MO ; si g est, à la surface du globe, l'accélération due à la pesanteur, ou la force rapportée à l'unité de masse pour une distance au centre égale à R, à la distance $\mathrm{OM} = r$ la force sera réduite à $g \frac{r}{\mathrm{R}}$. Cette force, proportionnelle à OM, se décompose en deux : l'une, proportionnelle à MI, qui agit dans la direction du mouvement ; elle est égale à $\frac{gx}{\mathrm{R}}$; l'autre, normale à la trajectoire, est proportionnelle à OI ; elle est constante par conséquent, et mesure la pression normale exercée par le point sur la droite AB. Nous la représenterons par $\frac{ga}{\mathrm{R}}$, en appelant a la distance OI. Cette pression normale développe, dans le sens opposé au mouvement, un frottement constant égal à $\frac{fga}{\mathrm{R}}$. Comptons les x positivement de I vers A. Alors la force $\frac{gx}{\mathrm{R}}$ tendra à diminuer les x, et la force $\frac{fga}{\mathrm{R}}$ à les augmenter, tant que le parcours aura lieu dans le sens AB. L'équation du mouvement sera donc

$$\frac{d^2x}{dt^2} = -\frac{gx}{\mathrm{R}} + \frac{fga}{\mathrm{R}} = -\frac{g}{\mathrm{R}}(x - fa).$$

Pour intégrer cette équation, prenons pour variable la différence $x - fa = x'$, ce qui revient à déplacer l'origine I, dans le sens positif, de la quantité II' $= fa$. Il viendra, en remplaçant x par $x' + fa$, l'équation

$$\frac{d^2x'}{dt^2} = -\frac{g}{\mathrm{R}} x',$$

Figure 17.1. The first page of Collignon's 1882 *Problème de Méchanique* in which it is shown that a particle's motion through a chord-cut tunnel will be periodic.

"le mouvement d'un point pesant sur une corde de la sphère terrestre," and developed the equations showing that the travel time along any straight-line tunnel cut through the Earth's interior (viewed as a uniform density sphere) and orientated at any angle to Earth's polar axis would be the same as that derived for the diametrical case.[2] Sadly, Collignon provides no background motivation for his analysis and the paper contains no references to other earlier works.

To see how Collignon's result works, let the tunnel subtend some angle φ with respect to the Earth's center O, and with reference to Figure 17.2 we can resolve the force term $F(r)$ driving the particle along the tunnel in the y-direction.

The equation of motion for an object of mass m falling through the off-center tunnel at a distance r from O will be $ma = md^2y/dt^2 = -F(r)$ $\sin(\theta)$. Once again assuming the Earth has a constant density ρ, the force term $F(r)$ can be expressed as $F(r) = (4/3)\pi r^3 \rho Gm/r^2$, which is linear in r, and the equation of motion reduces to $md^2y/dt^2 = -F(r)\sin(\theta) = (4/3)\pi\rho Gmr\sin(\theta)$. By straightforward trigonometry, however, we also see that $\sin(\theta) = y/r$, and accordingly, $d^2y/dt^2 = -ky$, where $k = (4/3)\pi\rho G$. The final form of the equation of motion is that of simple harmonic motion, and

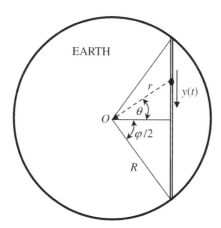

Figure 17.2. The geometry appropriate to Collignon's slant tunnel. O is the Earth's center and R is the Earth's radius.

---

[2] It is worth pointing out that Mr. S. A. Swann's answer to Cymro's question (recall Chapter 3 and see the Appendix) follows Collignon's method of solution.

the period of oscillation is $T = 2\pi(4\pi\rho G/3)^{\frac{1}{2}}$. Further substitution for the Earth's density in terms of the surface gravitational acceleration g can be incorporated, with $g = (4/3)\pi R\rho G$, giving us the result that $T = 2\pi(R/g)^{\frac{1}{2}}$. We note at this stage that the derived oscillation time is independent of the angle, φ, subtended by the tunnel at the Earth's center, and it also has the same form as that derived for the diametrical tunnel. Accordingly, the transit time between the ends of any Earth-crossing tunnel, whether it cuts through Earth's core or not, is always the same and equal to 42 minutes.

In addition to showing the equality of tunnel crossing times, Collignon went one very practical step further in his analysis and considered what the atmospheric pressure would be within the tunnel, and importantly, he found it to be totally overwhelming. Indeed, his calculations indicated that the atmospheric pressure inside even a *short-hop* tunnel passing just a few tens of kilometers under the Earth's surface would be many hundreds of times larger (see details below) than that experienced at Earth's surface — "C'est formidable."[3] Here is the practical engineer coming to the fore, and indeed, the atmospheric pressure within an open-ended, Earth-crossing tunnel would present an almost impenetrable barrier to any train trying to pass along it via gravity alone or even with the help of a powerful mechanical engine. One of the reasons behind this overwhelming damping effect is that of the drag force.

In pushing through a gas (or fluid) of density ρ, the drag force F_{drag} that an object experiences is typically expressed through Stokes' law (named after British mathematician Sir George Gabriel Stokes, 1st Baronet) which gives: $F_{drag} = \frac{1}{2} \rho C_D A V^2$, where C_D is the so-called drag coefficient, and where A and V are the cross-section area and the velocity of the falling object respectively. The drag coefficient varies in a complex manner according to the speed and shape profile, but in non-turbulent flow regimes its value is typically between $\frac{1}{2}$ and 1. Now, the larger the drag force, the greater is its decelerating effect upon the motion of a falling body. In the case of an open-ended Earth-tunnel, therefore, as the density increases with depth, so the larger and larger will be the retarding drag force. One can try to minimize the drag effect by say making the

---

[3] This was the exclamation made by Antoine Redier in his review article on Collignon's paper in the May 19th, 1883, issue of *La Nature*. See also [28].

falling object long and thin, thereby reducing the cross-section area *A*, but as we shall see below the density inside the Earth-tunnel increases so rapidly that any initial benefits are soon negated. The motion that results when drag forces are included in the Earth-tunnel problem are those of damped harmonic motion, that is, the oscillation amplitude about the Earth's center is steadily reduced over time (see Chapter 20). One can imagine the object oscillating backwards and forwards within the tunnel as described earlier (recall Figure 3.2), but now, the amplitude of motion decreases with each successive oscillation. This damping will continue until the body comes to a complete stop at the center of the tunnel — that is, the object will eventually come to rest, in a finite amount of time, at the center of the Earth.[4] There is a certain irony in this *real-world* result, in the sense that Aristotle and the Merton calculators (as discussed earlier) were right in their prediction that an object dropped into an Earth tunnel will, after a finite amount of time, come to rest at the Earth's center. Of course, it is important to point out that while these ancients may have been right in their prediction, they were right for the entirely wrong physical reasons.

Implicit in all of the calculations that have been presented thus far is the notion that the only force affecting the motion of the falling body is that of gravity. In an idealized thought experiment, this situation is fine, but in the real world the most common opposing force that a body encounters is that of air resistance. Édouard Collignon, as an engineer, well knew of this effect and this is what he attempted to evaluate in his 1882 paper. The problem is non-trivial. To begin, however, let us develop an approximation for the pressure variation with height within the Earth's atmosphere.

The equation of hydrostatic equilibrium $dP/dz = -g(z)\rho(z)$, where $g(z)$ is the gravitational acceleration and $\rho(z)$ is the density, provides a relationship between the pressure variation dP across a unit-area thin slice of

---

[4] The situation is actually much worse than just described, with the drag force becoming so large that any initial motion will be completely damped out long before the object actually reaches the Earth's center. Indeed, it will come to a hovering stop — just like the Rev. Mr. T-H-'s falling bullet (see the Appendix) — somewhere in the tunnel, with a balance being struck, at some interior point *r*, between the downward gravitational pressure force [*g(r) A*] and the upward acting air pressure force *P(r)* in the tunnel [29].

atmosphere of thickness dz. Assuming that the atmosphere is isothermal with constant temperature T and composed of an ideal gas of molecular mass μ, then $\rho(z) = P(z) (\mu/\Re)/T$, where μ is an appropriate mean molecular weight and $\Re$ is the gas constant. Accordingly, we have: $dP/P(z) = -g(z)[(\mu/\Re)/T]dz$. Under the approximation of constant gravity, $g(z) = g_0$, where g_0 is the acceleration at Earth's surface, and writing $H = [T(\Re/\mu)/g_0]$, which has units of distance and is accordingly called the atmospheric scale height, we obtain $dP/P(z) = -dz/H$. This expression can be integrated between $z = 0$, the Earth's surface, to some atmospheric height h, to yield an expression for the pressure as $P(h) = P_0 \exp(-h/H)$, where P_0 is the atmospheric pressure at Earth's surface ($P_0 = 10^5$ pascals). The negative sign in the derived expression for the pressure indicates an exponential decrease of pressure with height, with the pressure decreasing by a factor of $e^1 = 2.718...$ (the so-called e-folding distance) every H kilometers (for typical values in Earth's atmosphere, $H = 8.5$ km).

The various assumptions and constraints in the derivation just presented can generally be relaxed to provide a better atmospheric model, since the gravitational acceleration, the atmospheric composition and the temperature all vary with height. Here we will just relax the constant-gravity condition. In this case, $g(z) = g_0/(1 + z/R)^2$, where R is the Earth's radius, and the equation to solve for the pressure reduces to $dP/P(z) = -dz/(1 + z/R)^2 H$. Straightforward integration of this equation provides the pressure variation with height:

$$P(h) = P_0 \exp\left(-\frac{h}{H} \frac{1}{(1+h/R)} \right) \qquad (17.1)$$

Once again the pressure decreases exponentially with increasing height but with a slightly more complex e-folding distance than before.

The density at height h can be determined from the ideal gas equation, with $\rho(z) = P(z)/(g_0 H)$. With respect to my airplane journey from Calgary to London, Heathrow (recall Chapter 8), the cruising height was about 11 km altitude, and this indicates a characteristic atmospheric pressure of 2.75×10^4 pascals (or about $P_0/3.5$) and a density of 0.33 kg/m^3. These values are both a little high with respect to the U.S. Standard Atmosphere data, but remember our assumption in deriving Equation (17.1) is that the temperature is constant with height (it actually decreases by about 40 degrees between the Earth's surface and 11 km altitude). Nonetheless, we can now estimate the drag force experienced by the Boeing 777 aircraft that I was seated in. As we saw earlier, the drag force equation, $F_{drag} = \frac{1}{2}\,\rho C_D A V^2$, requires that we estimate the density, speed and cross-section area. The aircraft specifications given by Boeing indicate a cross-section area of $A \approx 113$ m² for the 777 aircraft (this is about the area of a circle of radius 6 meters). Furthermore, the drag coefficient for a typical subsonic aircraft is of order $C_D = 0.01$. With a flight speed of $V = 950$ km/hr $= 264$ m/s, the drag force experienced by the 777 aircraft at an altitude of 11 km is $F_{drag} = 13{,}000$ newtons. This drag force must be compensated for by the mechanical force of the aircraft's engines — the two General Electric GE90 engines on a Boeing 777 provide a combined forward trust of some 800,000 newtons, and this thrust easily overcomes the drag force, and it also enables the aircraft to lift off, carry and rapidly transport its total weight (aircraft, fuel, passengers and cargo) of some 300 metric tons.

While the above details are all important considerations for the flight of an aircraft, what is of more direct concern to the consideration of motion in an Earth-crossing tunnel or a very deep well is that akin to engine failure on an aircraft. Indeed, it is the attainment of a terminal free-fall velocity. The terminal velocity corresponds to the attainment of a maximum velocity as a body falls through a fluid or gas (such as the Earth's atmosphere). A maximum downward speed, the terminal velocity V_T, comes about when the drag force exactly matches that of the downward force of gravity. For an object of mass m falling in a region in which the gravitational acceleration g_0 can be taken as constant, Newton's second law gives the equation of motion as $F = mg_0 - F_{drag} = mg_0 - \frac{1}{2}\rho C_D A V^2$. When the downward force of gravity is in balance with the resisting drag force, so $F = 0$, and the terminal velocity attained will be:

$$V_T = \left(\frac{2mg_0}{\rho A C_D}\right)^{1/2} \tag{17.2}$$

Equation (17.2), for example, provides an explanation as to why the raindrops within a given downpour all hit the ground at the same speed — taking the typical size and mass of a raindrop to be 4 mm and 0.034 g respectively, then for a drag coefficient of $C_D = \frac{1}{2}$ (typical of that for a sphere), the terminal velocity will be of order $V_T = 9$ m/s (taking $g_0 = 9.81$ m/s^2 and $\rho = 1.29$ kg/m^3). A similar calculation for a golf ball (of diameter 4.2 cm and mass 46 g) gives $V_T = 32$ m/s. These latter two examples relate to motion through Earth's atmosphere towards the ground, to determine how drag will affect the motion of an object falling through an Earth-tunnel, however, we first need to make an adjustment to the pressure–height variation Equation (17.1). Since we will continue to assume a uniform density for the Earth, the gravitational acceleration experienced within an Earth-crossing tunnel will decrease linearly with depth, giving $g(h) = g_0(1 - h/R)$, and the pressure at a height h below the Earth's surface becomes:

$$P(h) = P_0 \exp\left(\frac{h}{H}\left(1 + \frac{h}{2R}\right)\right) \qquad (17.3)$$

Now, through Equation (17.3), we have an exponentially increasing pressure with depth underground. Indeed, the pressure is twice that at the Earth's surface ($2P_0$) at a depth of $h \approx 6$ km. At 10 km underground the pressure is $3.25P_0$; at 25 km underground, the pressure is $19P_0$; and at 50 km underground, the pressure is $367P_0$. Figure 17.3 shows the variation of pressure with depth as provided for by Equation (17.3). The atmospheric density at 11 km underground, the mirror height of my Boeing 777 flight, is about 4.4 kg/m^3, which with all else being the same, indicates a 13-fold increase in atmospheric drag.

Rather than pushing the bounds of our already stretched imaginations too far and consider the motion of a Boeing 777 aircraft flying down an Earth-crossing tunnel, let us consider something much smaller in scale — say a golf ball. We have already seen that a golf ball let fall through the atmosphere will attain a terminal velocity of 32 m/s, what now, we ask, is the motion and terminal velocity of a golf ball let fall through the entrance of an open-ended Earth-tunnel. The equation of motion for the golf ball will be the same as described earlier, with:

$$F = ma = mg - F_{drag} = mg - \tfrac{1}{2}\rho(h)C_D A V^2 \qquad (17.4)$$

where $a = d^2h/dt^2$ is the acceleration experienced by the golf ball, g is the gravitational acceleration $g(z) = g_0(1 - h/R)$, and where the atmospheric

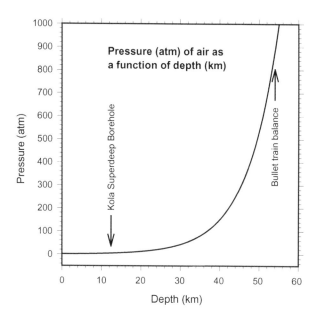

Figure 17.3. Atmospheric pressure versus depth inside an Earth tunnel (from Equation (17.3)). The pressure at the bottom of the Kola Superdeep borehole (Chapter 19) is indicated, along with the depth at which a Bullet Train would hover motionless in an open-ended Earth tunnel — see reference [29].

density $\rho(h)$ will vary in accordance with Equation (17.3) and the ideal gas equation. Equation (17.4) can only be solved for numerically [30]. With a Stokes' law drag variation we find that a golf ball will descend no more than 50 km below the Earth's surface before it is brought to a hovering stop after falling for some 95.2 minutes. The maximum velocity of the golf ball is also found to be a mere 34 m/s, and this velocity is achieved just 11 seconds after being released into the Earth-crossing tunnel. Replacing Stokes' law with a constant sliding friction term, with the golf ball just making contact with an otherwise evacuated tunnel (making the second term on the right-hand side of Equation (17.4) constant), still results in the golf ball coming to a gradual stop within the tunnel long before reaching the Earth's center. Indeed, this stopping condition is achieved for a golf ball if the sliding friction exceeds some 4 newtons per kilogram. Friction, either through air resistance or contact with the tunnel walls, is indeed an absolute killer of the Earth-crossing ideal.

In Figure 17.3 (shown above) the depth range has not extended beyond 60 km underground since the pressure at depths beyond this distance will be well beyond 1000 times that at Earth's surface, and at these kinds of pressures the ideal gas law will no longer hold true. Indeed, with increasing depth the gas molecules will be squeezed closer and closer together and it will no longer be legitimate to consider the gas as being composed of non-interacting point particles. A correction term derived by the Dutch physicist Johannes van de Waals could be applied to the equation linking together the pressure, density and temperature, but this would not help us for very long, since the pressure will soon pass into the domain where a phase change will occur and the gas will be transformed into a liquid. A nitrogen gas at room temperature, for example, will undergo a phase transition to a liquid once the pressure exceeds $34P_0$; it will undergo a phase transition to a solid, at room temperature, once the pressure exceeds some 20 GPa (or $2 \times 10^5 \, P_0$) — assuming that Equation (17.3) is still representative under these latter extreme conditions (which it isn't really), then the nitrogen gas within the tunnel will solidify at depths below about 100 km underground. A more detailed calculation would need to take into account the full thermodynamics of the situation, but there will be no escaping the inevitable result that at depths much deeper than about 50 to 150 kilometers below the Earth's surface any open-ended tunnel would be completely impassible. We have pushed our thought experiment too far, and by trying to turn it into reality the dictates of the physical world have stopped us solidly in our tracks.

In terms of practical engineering, Collignon's analysis indicates that if one is going to construct a transport tunnel between any two locations on the Earth's surface, then the tunnel would have to be sealed off at each end, and maintained in a near complete vacuum state, or, if the tunnel is open-ended, then it cannot be longer than about 1,600 km in length.[5] The former requirement, of course, would add extra engineering complications and costs to what would already be a formidable budget for the tunneling alone. The latter requirement greatly limits any potential tunnel network to be land-based, since no large ocean-spanning step can be accomplished by a single tunnel. In reality, there is no practical, cost-effective way of engineering

---

[5] At this chord length the tunnel will be no deeper than 50 km below the Earth's surface.

what have become known as gravity train system. After his 1882 analysis, Collignon abandoned any attempts to push forward the idea of gravity trains running through Earth's deep interior, and the subject essentially lay dormant within the scientific/engineering community for some 80 years. Interestingly, and perhaps appropriately, however, the idea of gravity trains was raised by Lewis Carroll in his novel *Sylvie and Bruno Concluded* published in 1893. In his dialog, Lewis introduces the character of Professor Mein Herr who explains that "each railway is in a long tunnel, perfectly straight; so of course the middle of it is nearer the center of the globe than the two ends; so every train runs half-way downhill, and that gives it force enough to run the other half up-hill." Lewis (that is Charles Dodgson) was a mathematics don at Oxford University and he may well have come across Collignon's work (or a news article related to it) soon after its first appearance in 1882. For all this, Lewis, through Professor Mein Herr, appears to have missed the point that the train journey time would always be 42 minutes.

In 1966, adhering to the rubric that "you can't keep a good idea down," Paul Cooper, at the Applied Research Laboratory of Sylvania Electronic Systems, in Massachusetts, revised the gravity train idea in a publication to the *American Journal of Physics* [31]. Cooper was looking to the future in his article and begins, "In his quest for intercontinental travel, man has probed the seas, rocketed through the air, and speaks now of traveling half across our planet in 40 min in a semisatellite or ballistic vehicle." With a further nod to Jules Verne's *Journey to the Center of the Earth*, Cooper then determines (without apparent knowledge of the fact) Collignon's slant condition and he further examines the circumstances for the fastest travel time profile for the tunnel (see Chapter 18). Interestingly, Cooper concludes his article by noting that the gravity train conditions apply not just to the Earth, but to the Moon and Mars as well. Apparently missing the point about atmospheric drag, Cooper suggests that trans-lunar and trans-Mars transportation systems will not suffer from the problem of earthquakes.[6]

---

[6] This, in fact, is not actually true in the sense that both lunar and Martian quakes have been recorded by surface landers at various times. Additionally, while the Moon has a solid core, that of Mars may still be partially molten, although it would appear that surface volcanism on Mars has now ceased. The Moon, of course, has no atmosphere, and that of Mars is significantly less extended than that on Earth.

It is probably safe to say that no realistic cost/benefit analysis is likely to support the utility of constructing even a relatively short distance gravity train systems — there are just too many alternative and cheaper ways of traveling to make such a scheme fiscally viable. This may not be the case in space (in the relatively near future), where one might well find value (both in the sense of money and utility) in mining and tunneling completely through large asteroids and perhaps the smaller planetary moons (see Chapter 22).

Chapter 18

Fastest Descent

And the fifth angel sounded, and I saw a star fall from heaven unto the earth:
and to him was given the key to the bottomless pit...

Revelation 9:1

Collignon's slant solution to the Earth-tunneling problem tells us that the travel time of a gravity train, between any two locations on Earth's surface, is always 42 minutes. This indicates an isochronal condition similar to that attached to the classical tautochrone problem. In the latter case, however, the travel time is independent of the starting position — in the gravity-train situation this is only true if the train starts at the Earth's surface; if it starts from somewhere inside the tunnel the travel time to its end point will no longer be 42 minutes. Not only is there a similarity between the Earth-crossing problem and the tautochrone problem, the Earth-crossing problem has some additionally verisimilitude to the brachistochrone problem. In the latter case it is the path of fastest travel that is being sought, and this may not necessarily be a straight line. Collignon's solution to the Earth-crossing problem entails making a straight-line tunnel, what if, however, a different path, indeed a curved tunnel, is employed instead — can this reduce the travel time, and if so, by how much?

The tautochrone and brachistochrone problems have a distinct and important place in the history of mathematics, and they introduce two additional problems that have become right-of-passage classics for present-day students of mathematics and physics. The first of these problems asks what is the shape of the curve for which the time taken, under the action of gravity alone, for a body to travel between two points A and B (with A above B, and B being the curve's lowest point) that is independent of the starting point A? The word tautochrone is based upon the ancient Greek for *same time*, and the problem is effectively asking what is the curve traced out by a truly isochronal (equal time) pendulum — a pendulum which has the same period of oscillation irrespective of its starting off-set angle. The second problem has an origin that can be traced back to June 1696, and it relates to a mathematical challenge thrown down to the "most brilliant mathematicians in the world" by Johann Bernoulli. Indeed, Bernoulli, asks, "given two points A and B in a vertical plane, what is the curve traced out by a point, moving under the influence of gravity only, which starts at A and reaches B in the shortest time." This is the brachistochrone, meaning *shortest time*, problem. The two curves that satisfy the equal time of descent and the fastest time of descent need not, in general, be the same, but it transpires that in a uniform gravitational field (such as that at the Earth's surface) they are the same curve, and the curve is the cycloid.

The word cycloid, meaning *circle-like*, was coined by Galileo in 1599, and it was Galileo as a young professor of mathematics at Padua who reasoned that of all the arch shapes that can span a given opening, that corresponding to the cycloid arch has the maximum load bearing properties. As a distinct curve, the cycloid was first studied by German philosopher Nicholas of Cusa in the mid-15th century, and it is defined as being the curve traced out by a point on the rim of a circular wheel as the wheel rolls along a straight line without slipping. The definition just provided, however, was only later presented by Marin Mersenne in 1623. During the later part of the 17th century the cycloid was in high vogue among mathematicians, and many remarkable results were determined with respect to its geometric properties. French theologian and mathematician Blaise Pascal was one of the early inspired[1] investigators of the cycloid, and in 1658 he not only published a text (under the pseudonym

---

[1] Perhaps inspired is the wrong term to use here since Pascal began to investigate the properties of the cycloid as a means of taking his mind off a raging toothache. Apparently, the cure worked.

of Amos Dettonville) about its properties, *L'Histoire de la Roulette*,[2] but he also offered prize money (to the sum of 60 Spanish doubloons) to anyone who could solve several of the outstanding problems that he had encountered. The prize money was to be split 40 to 20 doubloons to the first two correspondents that correctly determined the area and the center of gravity under a cycloidal curve. The time allotted to complete the challenge was 5 months, but there were no successful solutions within the set time limit and Pascal kept his money. For all this, many mathematicians were interested in the cycloid. Astronomer and architect Christopher Wren, for example, sent Pascal in 1658 his result for the rectification of the cycloid — this intriguing result indicates that the length of one complete arc of a cycloid is equal to the perimeter of the square circumscribed about the cycloids generating circle. Wren also developed a graphical method incorporating the rectification of the cycloid to solve for Kepler's Problem and the determination of a planet's position as a function of time in its orbit [32].

In an attempt to solve the problem of finding longitude at sea, Christiaan Huygens circa 1656 cautiously invented and then secretively had constructed the first pendulum regulated clock. His use of the (now iconic) pendulum was inspired by a reading of Galileo's 1638 *Dialogue Concerning Two New Sciences*, where it was argued that the swing of the pendulum is isochronal — that is the period of swing is independent of the starting off-set angle. Huygens found that Galileo's assertion was not, in fact, true, and accordingly in 1659 developed a new drive system in which the motion of a pendulum was forced to swing between two cycloidal cheeks[3] — that is, the length of the pendulum was continuously shortened and then lengthened in a smooth manner so as to remove the circular-arc error inherent in a fixed-length pendulum. Huygens found that the isochronal arc was that traced out by a cycloid through a hands-on, trial-and-error, graphical method, and it was not until 1686 before Gottfried Leibniz could show mathematically that the cycloid solved both

---

[2] Technically a roulette is the curve generated by a point attached to one curve as it rolls along another.

[3] The details behind Huygens' design were not made generally clear until 1673, at which time he published his important and influential text, *Horologium Oscillatorium*.

the tautochrone problem (see the Appendix) and the isochronal pendulum issue. As revealed in the Appendix, the travel time T_{cyc} along a cycloid-shaped wire under the influence of gravity is such that $T_{cyc} = 2\pi(a/g)^{1/2}$, where a is the radius of the cycloid-generating wheel. This result has the wonderful (thought experiment) consequence that we can think of the entire Earth, having radius R, being rolled underneath some cosmic straight-edge to trace out an inverted, giant (world-line[4]) cycloid. The travel time for a frictionless bead moving along our imagined world-line cycloid will be $T_{cyc} = 2\pi(R/g)^{1/2}$, which is the same as the oscillation period for a particle moving inside of a straight-line Earth-tunnel under constant gravity[5] (Figure 18.1).

The fastest descent, brachistochrone, problem was introduced by way of a public challenge in 1696. The problem, as first explained by Swiss mathematician Johann Bernoulli in the journal *Acta Eruditorum*, runs as follows, "given two points A and B in a vertical plane, what is the curve traced out by a point acted on only by gravity, which starts at A and reaches B in the shortest time." Bernoulli's challenge was addressed to "the most brilliant mathematicians in the world," and indeed, this was the very group that responded. Correct solutions, indicating that the path of fastest descent was that traced out by a cycloid, were obtained by Isaac Newton,[6] Gottfried Leibniz, Jacob Bernoulli (Johann's older brother), and Guillaume

Figure 18.1. The oscillation period of a particle moving in a pole-to-pole Earth-tunnel is the same as the isochronal period associated with the Earth-generated cycloid.

---

[4] The term world-line is not being used here in the general relativistic sense, but simply means the path traced out by a point upon the surface of the Earth as it moves along our imagined "cosmic straight-edge."

[5] One somehow feels that this result would have made a good mathematics Tripos question.

[6] Newton submitted his solution anonymously, but Bernoulli wasn't fooled, commenting, "ex unge leonem" — I recognize "the lion [that is Newton] by its claws."

de l'Hopital. The tautochrone and brachistochrone problems proved to be of immense importance in the development of new mathematical tools and methodologies, and they are now used as routine thought-experiment introductions to the study of the calculus of variation.

The time of descent along a brachistochrone curve, from its highest to its lowest point in a constant gravitational field, is the same as that for the tautochrone, namely: $T_{brach} = \pi(a/g)^{1/2}$, where a is the radius of the cycloid-generating circle. This result applies according to the generating circle rolling along a straight edge, but, one may additionally ask: "What is the travel time if the generating circle rolls along the circumference of a larger circle of radius R"? When the generating circle rolls along the circumference within the interior of the larger circle a hypocycloid is developed, and it is through such curves that the fastest travel time of a gravity train can be investigated. Numerous research papers [33] have been published concerning the brachistochrone problem under arbitrary potential fields such as $V(r) = kr^n$, where k and n are constants, but it transpires that the tautochrone and brachistochrone have the same solution curves (a cycloid) when $n = 2$ — for all other values of $n \neq 0$, they have different solution curves. Of interest to us is the fact that the $n = 2$ situation applies to motion within a uniform-density sphere — such as that exhibited by a gravity train. In the situation of uniform density, recall, the centrally acting force $F = -dV/dr = -2kr$ varies linearly with the distance r. In similar vein to the cycloid being the solution curve to the brachistochrone when the generating circle rolls along a straight edge, the solution curve for the brachistorone in the context of a uniform-density sphere is that of a hypocycloid (see Figure 18.2).

If the generating circle has a radius a and the larger uniform-density sphere has a radius R then the time T_B to travel along the brachistochrone arc is: $T_B = 2\pi[(a/R)(R - a)/g]^{1/2}$, where $g = GM/R^2$, with M being the mass of the uniform-density sphere. With reference to Figure 18.2, the ground path distance between the starting A and end point B of the hypocycloid is $s = R\theta_{AB}$, where θ_{AB} is expressed in radians, and $\theta_{AB} = 2\pi(a/R)$. The straight-line tunnel distance L between points A and B is $L = 2R \sin[\theta_{AB}/2]$, and the hypocycloid arc length S is: $S = 8(R - a)(a/R) \sin^2[R\theta_{AB}/4a]$.

By way of an example application of these results, let us return to my 2017 airplane flight from Calgary to London, Heathrow as introduced in Chapter 8 and illustrated in Figure 8.1. The angle subtended at the center

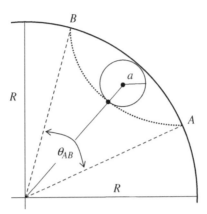

Figure 18.2. Hypocycloid generated by the small circle of radius a rolling along the inside of a great circle arc of radius R. The path takes the imagined gravity train from point A to point B, a distance s apart along the surface of the sphere.

of the Earth along the great circle path linking Calgary and London, Heathrow (a distance of some 7,023 km) is $\theta_{CL} \approx 63$ degrees $= 1.1$ radians, where the CL subscript indicates Calgary to London. Given $a/R = \theta_{CL}/2\pi$ and the Earth's radius is $R = 6,371$ km, we have that the radius of the hypocycloid-generating circle is $a = 1,118$ km, that the straight-line tunnel length is some $L = 6,725$ km, and that the cycloid-tunnel arc length is $S = 7,373$ km. The time to travel between Calgary and London is 42 minutes by the straight-line tunnel, but just 32 minutes by the brachistochrone path. Remarkable, while the hypocycloid path is some 650 km longer in length, the travel time from Calgary to London along the brachistochrone path is some 10 minutes shorter than that along the straight-line tunnel. While my airplane flight took some 8 hours to complete, traveling at a characteristic speed close to 950 kilometers per hour, the brachistorchrone traversing gravity train, at its maximum speed, would be moving at some 7.8 kilometers per second.

Taking one more look at the expression for the hypocycloid (that is brachistochrone) travel time within a uniform-density sphere: $T_B = 2\pi[(a/R)(R - a)/g]^{\frac{1}{2}}$, we note that in the special case that $a/R = \frac{1}{2}$, with the radius of the generating circle being exactly half that of the sphere within which it is rolling, the hypocycloid becomes a straight line, and the travel time $T_B = \pi(R/g)^{\frac{1}{2}} = t_{tunnel} = 42.2$ minutes. Any center-crossing straight-line Earth-tunnel, therefore, is a brachistochrone. A tunnel following Collignon's slanting path, however, may well be the shortest path between two points, but the motion is neither the fastest nor is it isochronal.

The solution to Cymro's problem, as illustrated in Figure 3.2 and solved for in the Appendix, is predicated on the Earth being a uniform-density sphere. Likewise, the brachistochrone calculation for the fastest crossing time uses this same idea. The question now is, is such a constancy realistic, and the answer to this is a very definite no. Indeed, the Earth's density increases by a factor in excess of 4 over its surface value (about 3,000 kg/m^3) to that at the center (about 13,000 kg/m^3). In the modern era the variation in the Earth's interior density is described by the Preliminary Reference Earth Model (PREM). Developed by Adam Dziewonski and Donald Anderson in the early 1980s, under the auspices of the International Association of Geodesy, the PREM is based upon a vast catalog of seismologic data. The model essentially provides a reverse-engineered profile to the Earth's density variations based upon the measured propagation times of sound waves produced by earthquakes [34]. The density profile derived in the PREM further informs us about the acceleration due to gravity within the Earth's interior. Indeed, the deduced variation of the density and gravitational acceleration, as a function of depth within the Earth, are shown in Figure 18.3. A distinct jump in the density is evident at the core–mantle boundary after which point the composition becomes increasingly iron rich — the liquid metal outer core extends between $0.19 < r/R_{Earth} < 0.55$ (recall Figure 3.3). The variation in the gravitational acceleration is further revealed by the PREM to be surprisingly constant within the outer 50% of the Earth by radius; decreasing in a near linear fashion thereafter to zero at the Earth's center.

The PREM-deduced behavior of the density and gravity variation within Earth's interior is important, and it requires that we re-interpret the whole physical basis of the Earth-tunneling problem. Let us back track a little. The trick to solving Cymro's problem and the pedagogical value of the Earth-tunneling question is entirely built around the constancy of Earth's density. The reasons for this assumption are clear enough, and it allows for an analytic solution to the equation of motion to be developed and it provides for an example of simple harmonic

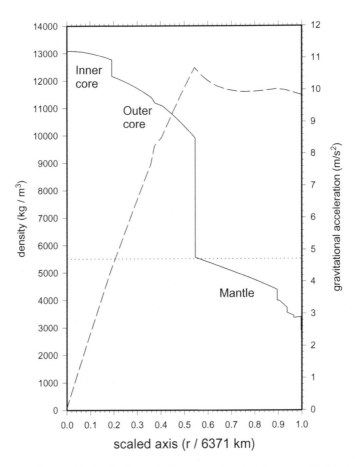

Figure 18.3. The Earth's density (solid line) and acceleration due to gravity (dashed line) variation as predicted by the Preliminary Reference Earth Model (PREM) [34]. The regions corresponding to the mantle, outer core and inner core are indicated in the diagram (recall Figure 3.3). The horizontal dotted line corresponds to the Earth's mean density of 5,514 kg/m^3.

motion to be found. There is nothing specifically wrong with assuming a uniform density in any model of the Earth, as long as one does not then try to invoke real-world results. The PREM results indicate, however, that in terms of physical motivation, the Earth-tunneling problem is better predicated on the assumption of constant gravity rather than that of constant density. That is, the straightforward approximation given by Galileo in his 1638 *Dialogue Concerning Two New Sciences* (recall Chapter 12) provides the best model approximation. To illustrate this point the equation of motion $d^2r/dt^2 = -g(r)$ has been numerically integrated with $g(r)$ being taken from the PREM data. We do not encounter Euler's "deplorable infinity" in this integration (recall Chapter 16) since the value of the acceleration at the Earth's core in the PREM case is zero, $g(0) = 0$, and not infinity. The PREM integration analysis reveals that the travel time along a tunnel to the Earth's center is 18 minutes 8 seconds, with the center-crossing speed being 9.8 km/s. This result is remarkably similar to the constant-gravity calculation described by Galileo, for which the travel time to the Earth's center is 18 minutes with the center-crossing speed being 11.2 km/s. In contrast, the constant-density model provides a center-crossing time of 21 minutes 6 seconds and a center-crossing speed of 7.9 km/s. The reason why the constant-gravity approximation compares so well with the PREM integrated result is simply that the gravitational acceleration is essentially constant throughout the outer half of the Earth's interior (as seen in Figure 18.3), and that by the mid-way point the falling body has already achieved most of its final speed. The PREM integration calculations reveal, in fact, that at the mid-way point through the Earth's interior an object will have achieved a velocity of 8 km/s, which is some 82% of its maximum center-crossing velocity.

Chapter 19

The Kola Pin-Prick and the Iron Blob

When you go down a coal-mine it is important to try to get to the coal face....
When the machines are roaring and the air is black with coal dust [...] the
place is like hell [...] heat, noise, confusion, darkness, foul air, and, above
all, unbearable cramped space.

George Orwell – *The Road to Wigan Pier* (1937).

Given the sublime beauty of the landscape on Vancouver Island it
seemed a little ironic, if not sacrilegious, to visit and then descend into
the Horne Lake Caverns. Nonetheless, *indoorsman* that I am, this is
exactly what happened one gloriously sunny day in August 2017. The
view from the visitor's center was veiled and faintly browned from forest-
fire smoke blowing in from mainland British Columbia. The day, however,
in spite of the haze, was bright, humid and sticky. To reach the cave
entrance our guide took us up a steep forested path, the sunlight marking
our way with dappled lakes of illumination. As we ascended through the
trail our breath quickened and sweat began to pool on our brows. It was
a magical landscape and as we approached the cave entrance, through a
narrowing tree-lined gully, there was a faint sense that we might easily
chance to catch, between the tree trunks and sunlight curtains, a fleeting
glimpse of magical beasts and questing knights. The cave entrance itself
was altogether different — it was dark and slightly perturbing, oozing
mystery and steeped in shadow.

At the entrance our guide instructed us in the use of the three-
points-of-contact method of exploration — use two feet and one hand,
or one foot and two hands, at all times, and keep your head low as you

progress forwards. In synchronicity we turned on the lights attached to our hardhats: the beams seemed remarkably feeble in the sunlight. There is a metal grill that covers the cave entrance — prompting one to question, is it there to stop people from falling in, or is it there to stop something from coming out. Once the entrance to the grill is opened we descend a dozen narrow steps and the outside world immediately begins to regress. The sunlight falters and the temperature drops, sounds become muted — there is a sense of dampness. The cave floor is a jumble of smooth, water-warn boulders; above us, is a scarred and twisting roof spattered with glistening beads of light. The *cave glitter*, our guide explains, is the result of our head-lamps illuminating little spheres of water located at the ends of growing stalactite tubes. They give the cave a fairy-tale appearance, and remind me of the Milky Way and the myriad stars spattered upon the darkness of the celestial vault.

We scramble and drop deeper into the cave, and the ponderous mass of rock around us starts to take on a tactile feel — it is oppressive and confining; this is not a world made for human beings. Indeed, the cave reminds me of the great story of the Earth, with the ancient and the modern all mixed into one incredible moment. The cave cuts through limestone rock, composed of the dead animals which once lived in a long-ago lost sea. This same seabed, made of those once living animals, has been further thrust upwards, by the relentless power of plate tectonics, to make mountains and marvels. This same ancient seabed has also been run over, crushed and scoured by massive ice sheets, descending from the north, during the times of numerous ice ages. The present-day mountains of Vancouver Island have indeed taken a long, hard and tortuous path to expose their present magnificence. Chemistry begins the process of cave formation: rainwater being slightly acidic slowly dissolves the calcium from the limestone, and slowly, over many eons, it begins to carve out channels and groves. The channels grow and eventually, if the conditions are right, merge to produce clints and grikes, and baby caves. Slowly, but assuredly, the ancient rock, made from myriad ancient sea animals, pushed upwards by the force of tectonic motions, is riven with hollows and indentions. Once the caves are deep enough and sufficiently connected they can channel what was previously surface running water into a subterranean river. Now, driven underground, the stream of water can begin to carve out and enlarge, over thousands of years, the interior of the cave. It is along such an ancient stream-riven floor that we now walk.

The path twists and turns and our lamps can only illuminate a fraction of the cavern at any one time, the shadows drawing deep into the ceiling and walls — it is a very local world that enshrouds us. Along the cave walls and ceiling we see calcite gems. The calcium that the slightly acidic rainwater has removed from the limestone is now reforming, atom by atom, drip by drip. Slowly, at a rate of about 1 centimeter per century, the calcite grows to produce stalactites and stalagmites, drapes and tubes and spirals — it is entirely magical and marvelous. The calcite structures take on marvelous shapes and forms and our guide points out to us a howling wolf, an alligator with a cigarette, and a sitting Buddha statue — the mind works overtime to find familiarity in the alien landscape that is slowly growing about us.

The cave narrows as we descend deeper into its maw, and our motion changes from being upright to stooped — the cave is humbling us and we accordingly begin to bow. There is a sense that the cave could go on forever, becoming ever narrower — thrusting like a thin probing finger deeper and deeper into the Earth, searching out for some ancient and primordial inner energy. For a short while we turn off the lamps on our hardhats — the darkness is absolute and the silence is oppressive. It was like looking into the depths of an alien universe that harbored no stars. There is a palpable sense of relief when we turn our lights back on again — a cold, dark cave is no place for human beings to tarry. We head back to the entrance and as we climb out the light and heat of the day are welcoming and familiar — there is a sense of emerging from a slow and ancient place into a more dynamic and fecund world. And yet, for all this thrill of adventure, *indoorsman* that I am, we have descended no more than a few tens of meters below the surface rocks.

Whether caving as a tourist, as I did at Horne Lake Caverns, or working a coal seam, as described by George Orwell in his 1937 essay *The Road to Wigan Pier*, going underground is always a humbling and back-bending experience. Tunneling and the extraction of ore from deep mines, of course, have a long history, but such activities have been entirely dedicated to near-surface domains. To explore deeper, and to exercise scientific rather than corporate interests, one needs to drill downwards, aiming indeed for Earth's center. The problem with such geophysical research, however, is that Earth's core is located over 6,300 km below ground. Accepting, as we must, that Earth's core will never be directly drilled, and curtailing our goals to just probing the Earth's deep mantle, to engage with the Mohorovičić discontinuity, well, even this has not been achieved. The first serious attempt to drill through the Earth's crust was begun in 1961, and it was funded by the American Science Foundation. Named Project Mohole, five deep boreholes were drilled from a floating platform located off the island of Guadalupe. The deepest borehole was sunk to a depth of 183 meters below the seafloor — itself overlain by 3.6 km of water. While the test drilling showed that the technology was capable of excavating very deep boreholes, large cost overruns prompted the US Congress to cut the program's funding, and the project was shut down in 1966. Later, starting in late November 1972, the Lone Star Producing Company worked the Bertha Rogers oil-exploration borehole, in Oklahoma, to a depth of 9.583 kilometers (taking over a year and a half to complete the project). The German Federal Ministry of Research funded a deep drilling scientific program through the late 1980s to mid-1990s, with the KTB borehole, near Windischeschenbach in Bavaria, being driven to a depth of 9.101 kilometers.

The Kola Superdeep Borehole SG-3 is located about 50 km south of the Barents Sea on the northern rim of the Baltic Shield. The SG-3 borehole extends to a depth of 12.262 kilometers and it is the deepest geophysical borehole that has ever been logged and excavated. A product of the Cold War, the Superdeep program was initiated at the same time as the American–Soviet space race — with all the same bravado and nationalism. The drill site was selected in 1965 and it is located in a region where extensive copper and nickel mining has taken place. Drilling began on May 24th, 1970, and the 23-cm diameter borehole was pushed to its final depth by 1989. SG-3 was hollowed out by a then newly invented technique called turbo-drilling — in this manner the cutting bit is rotated through a turbine action by pressurized mud pumped down the pipe stack. The drill pipe is made of light-weight aluminum and the drilling rig (housed in a massive 200-foot tall building — see Figure 19.1) could make 2 to 3 meters headway per hour. Remarkably, the entire 12-km pipe stack could be retrieved and reinserted in less than a day. Drilling ceased in

Figure 19.1. The bleak landscape of the Kola Superdeep Borehole complex. The drilling rig was contained within the tall, 200-foot high, building seen at center right.

1992 due to higher-than-expected temperatures of some 180°C being encountered — at such temperatures the drill-head efficiency is greatly compromised. As the substrate temperature increases not only does drilling become more problematic, the lithostatic pressure due to the surrounding rocks will tend to collapse and close up the borehole.

Funding was eventually cut to the Superdeep project in 2006 and the once mighty drill site now lies in utter ruin having been completely abandoned in 2008. While the ending of the Superdeep program was somewhat unglamorous, it was, during its operational years, a major engineering and geophysical exploration triumph. The program provided fundamental data on deep-rock characteristics in a domain spanning the Proterozoic eon (extending from 2,500 to 541 million years ago) to the Archean basement (composed of gneissic granite over 2.5 billion years old). Perhaps one of the most remarkable results of the drilling program was that of finding abundant microfossils at depths amounting to some 6 kilometers underground. The core samples also revealed that at depths of some 7 kilometers below the surface the strata was highly fragmented and saturated with highly mineralized water — this water being derived, in fact, from deep-crust minerals rather than having percolated downwards from the surface. The core samples also revealed that an anomalous change in the propagation speed of seismic (sound) waves that had previously been recorded at a depth of about 7 kilometers below ground was due to a metamorphic transition in the granite present, rather than being due, as was initially assumed, to a transition discontinuity with the granite substrate giving way to basalt.

While other super deep boreholes have been drilled in other parts of the world, SG-3 still holds the depth record. In terms of the longest borehole, however, this record goes to the Sakhalin-1 oil and gas pipeline located on Sakhalin Island off the far northeastern coast of Russia. This particular borehole extends to a length of 12.345 kilometers and cuts to a depth of 11.475 kilometers. Drilling, in spite of a long history of technological developments, remains to this day a very expensive and relatively slow endeavor. For all this, the SG-3 borehole hardly made its way through about 1/3rd of the Baltic continental crust (estimated to be about 35 kilometers deep). Indeed, if the 12.262 km depth of the Kola Superdeep Borehole is taken to be the thickness of a single page of this book, about 0.1 mm, then the book-of-the-Earth will be some 52 mm across — indeed, a formidable tome. Drilling is not only expensive, it cannot take you very far in the direct exploration of the Earth's deep interior. To go deeper alternative methods will have to be found, and one such suggested approach is to create a real-world *China Syndrome* using a massive blob of molten iron to transport a probe all the way to the Earth's core.

The China Syndrome is a 1979 Hollywood film directed by James Bridges and starring Jane Fonder. The film's premise tells the story of safety cover-ups at a nuclear power plant, and it specifically considers the possibility of nuclear meltdown. Here the idea is that the high-temperature reactor-core material[1] might literally melt its way through any containment basement and leak catastrophically into the surrounding environment — melting "all the way from America to China"[2] (Figure 19.2). As a remarkable historical coincidence, the film went into public release just 12

---

[1] Such a molten mass of fissile fuel, concrete and debris is called corium lava.

[2] This, of course, is a classic Hollywood inaccuracy in that any wayward nuclear core released from America would descend to the Earth's core and not across the Earth to China.

Figure 19.2. Corium lava (formed of uranium dioxide fuel, graphite and molten concrete) burnt through the reactor basement floor during the Chernobyl (1986) nuclear power plant disaster. This image shows corium lava oozing from a high-pressure steam valve. The corium lava at Chernobyl had an initial temperature of about 2,200°C (easily high enough to melt through most surface rock types) and it remained above 1,600°C for nearly 4 days. According to the International Nuclear and Radiation Event Scale (INES), the Three Mile Island incident (where a small amount of corium lava did form) was rated as a level 5 "accident with wider consequences." Fortunately the Three Mile Island radiation leak remains the highest INES rated accident in the USA to this day; even so, it took 14 years to "clean up" the contamination. In more recent times the Three Mile Island incident has been largely overshadowed by the INES (maximum) level 7 "major accident" events at Chernobyl and at the Fukushima (Japan, 2011) nuclear power plant.

days before the Three Mile Island nuclear disaster in Pennsylvania. The film was largely concerned with managerial corruption and accident cover-up issues, but the core meltdown and "escape" scenario is real enough, and indeed, beginning in the early 1970s the idea of disposing of nuclear waste by self-burial was actively studied. Indeed, the deep underground burial of radioactive waste has a long and controversial history, but the basic concept is to store the waste out of harm's way, for many thousands of years, in deep underground bunkers. Self-burial takes this idea one step further and removes the requirement of actually drilling deep containment wells (or excavating deep bunkers) by allowing the waste to literally make its own self-sealing borehole. The self-burial process proceeds through the contact melting of underlying rocks by the high-temperature radioactive waste. As long as the nuclear waste can heat and melt the rock beneath it, it will sink downwards. Importantly, cooling in the region above the capsule will allow the substrate rock to recrystallize, thereby sealing in the waste material. The concept of self-burial is unnervingly simple, although fraught with complications. Indeed, the self-burial process can be modeled in terms of the drag (that is buoyancy) experienced by a hot rigid sphere (the capsule containing the nuclear waste) as it melts its way through a cold medium (the surrounding rock). Steven Emerman and Donald Turcotte (both of Cornell University) considered just this basic scenario in a ground-breaking 1983 research paper [35] entitled *Stokes's Problem with Melting*. Publishing their work in the *International Journal of Heat and Mass Transfer*, the authors concluded that a China Syndrome-like, *escaped* nuclear reactor core could melt its way to the Earth's core in about 2,000 years.

A self-sinking capsule, for geophysical research, capable of returning *in situ* data on the Earth's mantle and lithosphere, could be built around the idea of dissipative heating. In this situation a hot spherical capsule of temperature T_B, radius d and density ρ_B is envisioned to be falling through a medium of density ρ_m. Provided T_B is greater than the melting point temperature T_{melt} of the surrounding substrate, the capsule will gradually fall through the melted surroundings at some velocity u_0. The power, that is the rate of doing work, in melting the column of material directly below the capsule is $P_m = \pi d^2 u_0 \rho_m L$, where L is the sum of the latent heat of fusion required to melt the substrate medium L_m, as well as the heat required to increase the temperature of the surrounding substrate to its melting point: $L = L_m + c_p(T_B - T_{melt})$, where c_p is the specific heat capacity of the substrate material. Furthermore, the power P_D produced by dissipative heating is equal to the rate of the loss of gravitational potential energy of the capsule and this will be related to the buoyancy (drag) force acting on the capsule: $P_D = (4/3)\pi d^3 g(\rho_B - \rho_m)$. Equating the power produced by viscous dissipation to the power required to melt the substrate material gives a critical radius for the sphere. Namely,

$$d_{crit} = \frac{3}{4} \frac{\rho_m L}{g(\rho_B - \rho_m)} \qquad (19.1)$$

If the radius of the sphere $d > d_{crit}$ then the heat produced by viscous dissipation exceeds the melting point temperature of the substrate material, and the sphere will sink. Additionally, for $d > d_{crit}$ the temperature of the sphere will increase (through the release of gravitational potential energy) as it falls and so too will its velocity, with a thermal runaway effect coming into effect. If $d < d_{crit}$ then the capsule will stop sinking. The critical radius can be evaluated through the input of typical values, which are $(\rho_B - \rho_m) = 4,000$ kg/m³, $L = 4 \times 10^5$ J/kg, and $g = 9.8$ m/s². With these characteristic numbers we find that $d_{crit} = 30.5$ km — implying that for this mechanism to work a formidable sized sphere is required. There is a way to greatly reduce the size of the capsule, however, and this requires that it have some internal heat source (Figure 19.3). In this case, the internal heat source raises the temperature of the capsule's outer layer above that of the melting point of the surrounding material (this is effectively the self-burial condition invoked for the disposal of nuclear waste — *The China Syndrome* under semi-controlled conditions).

In the case of a spherical capsule heated by some active radionuclide [36], the total power of heat released Q will depend upon the amount of radionuclide material M and upon the specific heat released Q_R by the radionuclide: $Q = MQ_R$. The heat released by a capsule of radius d will

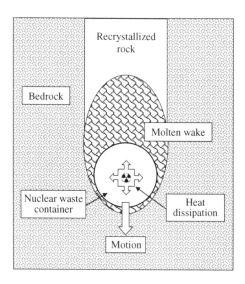

Figure 19.3.   The self-burial of radioactive waste through deep-rock melting.

accordingly be $q = MQ_R/[(4/3)\pi d^3]$. The condition for sinking now becomes $q > q_{th}$, where $q_{th} = 3\chi(T_{melt} - T_{rock})/d^2$, and where χ and T_{rock} are the thermal conductivity and temperature of the surrounding substrate. The sinking condition sets, for a given container radius d, a constraint upon the mass and the heat that must be released per kilogram by the radionuclide such that $MQ_R > 4\pi d\chi(T_{melt} - T_{rock})$. Taking a granite substrate by way of an example calculation, the thermal conductivity is of order 2 W/mK, the melting point temperature is $T_{melt} \approx 1{,}230$ K, and for a 1-m diameter container, so $MQ_R > 12.0$ kW, taking $T_{rock} = 283$ K. For cobalt-60, $Q_R = 17.4$ kW/kg, and accordingly some 700 grams of this radionuclide would be required to *drive* a self-sinking 1-m diameter sphere. The initial descent speed U_0 is estimated [36] to be of order $U_0 = (4d\rho_c q)/(3\rho_m L)$, where ρ_c is the density of the capsule (containing the radionuclide), and as before L is the sum of the latent heat of fusion required to melt the substrate medium L_m, as well as the heat required to increase the temperature of the surrounding substrate to its melting point: $L = L_m + c_p(T_B - T_{melt})$. If we place, say, 5 kg of ^{60}Co inside of a 1-m diameter capsule, then the initial total thermal power will be of order 87 kW, and the initial thermal power available to the capsule for melting its way through the rock substrate will be $q \sim 174$ kW/m^3. The initial speed will additionally be $U_0 \sim 7 \times 10^{-5}$ m/s or some 2 km per year. Given that the thermal power will decrease exponentially with time, such that at time t we have $q(t) = q(0)\exp(-\lambda t)$, where λ is the decay constant appropriate to the radionuclide. Given the condition for the capsule to keep sinking is $q(t) > q_{th}$, the capsule will keep sinking until t_{end} where,

$$t_{end} = \frac{1}{\lambda}\ln\left[\frac{q(0)d^2}{3\chi(T_{melt} - T_{rock})}\right] \tag{19.2}$$

The decay constant for ^{60}Co is 4.17×10^{-9} s^{-1}, corresponding to a half-life of 5.27 years. Accordingly, our example capsule will sink for a total of about $t_{end} \sim 15$ years. The approximate depth to which the capsule will sink is therefore of order 30 km. By placing more ^{60}Co into the capsule, the longer will the $q(t) > q_{th}$ condition hold, and the deeper will the capsule descend into Earth's mantle. Given ^{60}Co has a density of 8,860 kg/m^3, so our 1-m diameter sphere could in principle contain as much as 4,400 kg of this particular radionuclide, and under these circumstances the descent time will be increased by a factor of about 4.5 to some 70 years, resulting in a sink distance of about 140 km.

With self-burning, radionuclide-powered capsules it would be possible to explore the Earth's mantle to perhaps a few hundred kilometers, some 20 times deeper than the Kola Superdeep Borehole, on a timescale of perhaps half a century. While substantial depths of the Earth's mantle could be studied with self-sinking capsules, we have still hardly penetrated to the Earth's inner core — even at 150-km depth, only 2% of the journey to the core has been completed. One recent proposal whereby the Earth's core might be reached in a reasonable timescale, indeed, on the order of weeks, is that proposed [37] by David Stevenson (California Institute of Technology, Pasadena). Stevenson's method is grand in scale and builds upon the self-burial paradigm. Rather than using nuclear heating, however, Stevenson suggests that the instrument-carrying body should be a large blob of liquid iron. Here the idea is to use the heat energy of the liquid iron to propagate a downward moving crack in the substrate rock. The iron, being fluid, will then flow into the crack created in front of it. A bootstrap method thereafter sets in, with the liquid iron flowing ever deeper into the continually opening substrate crack ahead of it. The material above the iron blob will eventually cool and recrystallize, thereby closing up the path behind it. Stevenson considers a vertical crack of width d, horizontal length $W \gg d$, and depth H. Provided that the vertical crack depth is sufficiently large then the propagation speed is essentially that of the liquid-iron flow velocity, which can be modeled according to Poiseuille's equation. In this case, the propagation speed is given by the relationship: $U_{blob} = [(\rho_{blob} - \rho_m)gd^{5/4}/(\rho_m\mu^{1/4})]^{4/7}$, where $\mu \sim 10^{-6}$ m^2/s is the kinematic viscosity of liquid iron. Characteristically, this formula indicates a propagation speed of order $U_{blob} \approx 25d^{5/7}$ m/s near to the Earth's surface where the acceleration due to gravity is $g \sim 10$ m/s^2. The shear modulus of the rock through which the crack is propagating will present a pressure of order 10^{11} (d/H) Pa, and this stress must be overcome by the pressure head of the liquid-iron blob $(\rho_{blob} - \rho_m)gH$. These two relationships determine a characteristic crack depth $H \sim 3 \times 10^5$ $[(d/g)/(\rho_{blob} - \rho_m)]^{1/2}$ m, or characteristically $H \sim 1.5d^{1/2}$ (km), and the associated pressure head for the liquid-iron blob is some 7×10^7 $d^{1/2}$ Pa. If we assume $H = 500$ m, then $d \sim 11$ cm; and taking $W = H$, the amount of liquid iron required to fill the crack will be of order $\rho_{blob} dHW \approx 2 \times 10^8$ kg — this corresponds to a spherical mass of liquid iron some 40 meters in diameter. In 2016 the world's industry produced some 1.6×10^{12} kilograms of steel, indicating that the material necessary to produce an Earth-penetrating iron blob amounts to about one hour's worth of world steel production. Taking the crack propagation speed to be comparable to the liquid-iron flow speed, then the iron blob will move downwards at a speed of about 5 m/s, and at this speed the entire distance to the Earth's core could be covered in about a fortnight. To produce the initial crack into which the liquid-iron blob would be poured, Stevenson suggests using a thermonuclear bomb. Entraining a neutrally buoyant probe (having a size about that of a grapefruit) within the liquid-iron mass, its progress downward could be monitored from the Earth's surface, and if the instrumentation could be extended to measure substrate temperature and composition, then the characteristics of the Earth's interior could be mapped out.

Stevenson's liquid-iron blob constitutes a thought experiment on the grandest scale, and it is not a wholly unbelievable project that is envisioned. Many technological barriers would need to be overcome, however, in order to make such a scheme work, but as Stevenson notes at the very end of his article [37], "This proposal is modest compared with the space program, and may seem unrealistic only because little effort has been devoted to it. The time has come for action."

Chapter 20

A Black Hole Falling

To see the World in a Grain of Sand
And a Heaven in a Wild Flower
Hold Infinity in the palm of your hand
And Eternity in an hour

William Blake

Leonhard Euler, as we saw in Chapter 16, realized that lurking within Newton's equation of gravitational attraction was a potential pathology — a singularity, at $r = 0$, where the equation *explodes* to infinity. In everyday, typical usage, however, this singularity is never manifest, since objects have a finite size and the separation between centers is always finite. In the realm of the very small, and under the circumstances of gravitational collapse, however, the singularity within Newton's theory can be realized. By placing enough mass in a small enough volume of space, gravity, the weakest of all the known fundamental forces, can crush both matter and spacetime out of all existence. Such emotionally perturbing objects are called black holes, and they represent one of nature's darkest secrets — what happens inside of a black hole, stays inside of a black hole, and we, as outside observers, will never know what takes place within their interiors.[1] For our narrative, however, it is not what goes on inside of a black hole that is important, but rather it is the fact that such objects do exist

---

[1] I write this in spite of claims to the contrary by various luminaries of physics — my point essentially being that there may well be a theoretical escape hatch for information, but the information is in no way actually *readable* in any real world sense.

and that they may exist on a scale that is useful for probing Earth's interior.

Black holes emerge as a phenomenon associated with gravitational collapse — and in many ways they are inherently simple objects. All of the mess associated with the singularity, where gravity apparently becomes infinitely strong and where matter is both ripped apart and then packed to an infinite density in a region of infinitely curved space, takes place within a region beyond any observations that we can make. The singularity, where physics as we presently know it ends, is hidden from our direct view by an event horizon (a spherical region characterized by the Schwarzschild radius). The singularity at the center of a black hole of mass M resides within a spherical cloak with a radius R_S, with $R_S = 2GM/c^2$. Since the gravitational constant G is very small and c^2 is very large, it is evident that the event horizon radius R_S is typically going to be small. For the Sun, composed of some 2×10^{30} kg of gaseous matter, the Schwarzschild radius is 3 kilometers; for the Earth, the Schwarzschild radius is just 9 millimeters. The compression, as such, is colossal and entirely unimaginable, and yet gravity has conspired to produce black holes [38]. A massive black hole, Sagittarius A*, having a mass of several million solar masses (and a Schwarzschild radius of some 22 million kilometers) resides at the center of our Milky Way galaxy, and a 15-solar-mass black hole (with a Schwarzschild radius of 44 km) orbits the star HD 226868 once every 5 ½ days (this is the Cygnus X-1 system). The existence of these and similar such black hole systems are betrayed indirectly — we don't *see* the black hole, rather it is the effect of the black hole on its surroundings that is observed. In principle interstellar space is teeming with black holes with masses from as small as 10^{12} kg to many tens of solar masses. Indeed, one of the many existential disasters that could befall humanity, all be it a highly unlikely scenario, is that of encountering a stellar-mass rogue black hole. Intriguingly, however, it is entirely possible that the Earth encounters mini black holes on a near daily basis — and we are largely none the wiser, and/or at risk of sudden death.

In the early 1970s Stephen Hawking, as a young research student at Cambridge University in England, first studied and then published a whole series of ground-breaking articles on the possible existence of miniature and primordial black holes. These objects were envisioned to have formed in the hellish conditions applicable to the initial *spark* of the

Big Bang when the universe burst into existence. Not only this, but, it was suggested, that some of the primordial black holes could have been created with masses as small as the Planck mass, $\sim 2 \times 10^{-8}$ kg, giving them event horizon diameters as small as $\sim 6 \times 10^{-18}$ m. Subsequent studies by Hawking and co-workers further revealed that as a consequence of quantum mechanical effects operating at the Schwarzschild boundary, black holes might additionally emit radiation (so-called Hawking radiation) into space and slowly decrease in mass. The lifetime of a black hole against evaporation is dependent upon the cube of its mass and accordingly, given the universe has a present age of some 13.8 billion years, only primordial black holes with masses larger than 10^{12} kg can exist in the present epoch. Such a black hole will have a Schwarzschild radius of about 1.5×10^{-13} meters, making it significantly smaller, in fact, than an atom, but larger than an atomic nucleus.[2] It is their minuscule size that makes the hunt for such remnant primordial black holes both challenging and frustrating. Importantly for our narrative, however, the extremely small size of any primordial black holes dictates that they would either pass straight through the Earth, or (much less likely) be captured into a stable (Earth interior) orbit as a consequence of an encounter. And, while the mass of the primordial black hole is extremely small in comparison to the mass of the Earth, remarkably, the effects of an interaction, either tunneling or orbiting, might just be measurable.

A number of physical mechanisms might betray the passage of a miniature black hole (having a mass of order 10^{12} kg) through the Earth. Iosif Khriplovich and co-workers in Russia [39], and Yang Luo (Princeton University) and colleagues, have specifically investigated the possibility of detecting primordial black hole encounters through seismic measurements [40]. Specifically, both research groups argue that during the passage of a mini black hole through the Earth it will mechanically deform, through its gravitational influence, the matter along and surrounding its path, exciting thereby sound waves as it goes. Since the typical speed with which any primordial black hole will encounter the Earth must be of order 200 km/s (that is, of order the Sun's orbital speed at its location away from the galactic center) so the excited sound waves will have a

---

[2] The sizes of atomic nuclei range from as small as 1.6×10^{-15} m to as large as 15×10^{-15} m. The size range of atoms varies from about 10^{-10} m to 5×10^{-10} m.

velocity greatly in excess of the sound speed through rock. Accordingly, the research teams suggest, an acoustic analogy of Cherenkov radiation will arise, and this should produce a unique seismic signal. Additionally, since the Earth-crossing time of the mini black hole will be about 1 minute, so its resultant seismic signal should be recorded all over the Earth's surface at essentially the same instant. Luo and co-workers suggest that the seismic energy deposited by a 10^{12} kg mass mini black hole could be as large as a magnitude 4 earthquake. To date no seismic signal suggestive of the Earth having encountered a mini black hole has been observed. While the expected encounter between the Earth and a mini black hole will be a single tunneling event, Boris Zhilyaev (NAS of Ukraine) has argued [41] that the capture of the black hole may occasionally take place. Under these circumstances the black hole will either begin to oscillate back and forth across the Earth's diameter or enter into an orbit about its core. While such capture events are highly unlikely, it is additionally possible that at some future date a mini black hole might be created by accident on Earth. Irrespective of how such a captured black hole might end up within the Earth's interior the question arises as to whether such a capture event will result in an existential crisis for humanity.

Science fiction author Larry Niven wrote a Hugo[3]-award-winning short story in 1975 called *The Hole Man* — its theme has a strong resonance with Zhilyaev's suggestion and the controversy that erupted over the start-up and first beam collisions at the Large Hardon Collider (LHC) located at CERN. Indeed, several international court cases were filed against CERN in an attempt to close the LHC before it had even launched a single proton down its 27-kilometer long storage ring. At issue was the formation of miniature black holes. Niven leads us into the scenario, and he opens his short story with the lines "One day Mars will be gone. Andrew Lear says that it will start with violent quakes, and end hours or days later, very suddenly. He ought to know. It's all his fault." Niven's story revolves around the discovery of an alien communication device during the human exploration of Mars. The device, it turns out, uses gravitational waves as the communicating medium and at its core is a mini black hole.

---

[3] This is the same Hugo Gernsback that we encountered in Chapter 9 concerning the *Scientific Adventures of Baron Munchausen*.

Unfortunately, Lear mistakenly turns off the black hole containment field and it begins to sink into the Martian interior: "its gravity is ferocious" writes Niven, "and its falling back and forth through the planet [indeed, it is a self-tunneling, Earth-crossing analog of Cymro's problem] sweeping up matter. The more it eats the bigger it gets [...] sooner or later [it] absorbs Mars. By then it'll be just less than a millimeter across — big enough to see." Substitute Earth for Mars and Andrew Lear for CERN and the similarity between stories is complete. The court cases brought against CERN were not unreasonably founded on a human emotional level, but they were based upon unreasonable physical arguments. It is true that the LHC, under some theories allowing for multi-dimensional space (having between 6 and 10 dimensions), might possibly produce miniature black holes, and, if so produced, such black holes would indeed sink to the center of the Earth. The point, however, is that the most massive mini black hole that might conceivably be produced in the LHC would have a mass of perhaps 10^{-23} kg, giving it an event horizon radius many orders of magnitude smaller than an atomic nucleus, and accordingly such a black hole would not be unable to accrete matter and grow [42] — additionally, the lifetime against destruction of such a black hole through Hawking radiation would be a fraction of a second.

While the end of the Earth through black hole consumption will not come about through the generation of mini black holes within the LHC, this does not mean that such an end fate is completely out of the question; if not for the Earth, then for some other hapless planet somewhere else within the galaxy. The black hole version of Cymro's problem, as described by Niven with respect to Mars, is described according to the equation for simple harmonic motion with an additional damping term accounting for the increase in the black holes mass. Accordingly, under the assumption of a constant-density Earth, the equation of motion for the self-tunneling black hole becomes

$$\frac{d^2r}{dt^2} + \left(\frac{dm/dt}{m}\right)\frac{dr}{dt} + \omega^2 r = 0 \qquad (20.1)$$

where, as we have seen before, $\omega = (4/3)\pi G\rho$, and where the additional velocity-dependent term accounts for the momentum loss due to mass

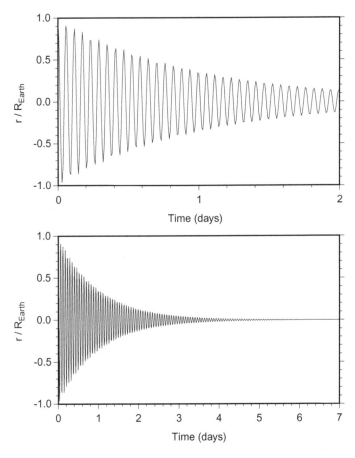

Figure 20.1. Damped oscillatory motion of an accreting black hole moving inside of the Earth. The solution to Equation (20.2) shown here adopts $\omega = 1.240 \times 10^{-3}$ (appropriate to a constant-density Earth model) and $\beta = 0.019$, giving an *e*-folding time of about 1 day, and a settling time of 7 days. At the end of the settling time the mass of the black hole will have increased by a factor of one million times. The top panel shows the first 2 days of motion, while the lower panel shows the motion over 7 days.

accretion. The (*dm/dt*) term accounts for the rate at which the black hole mass increases by swallowing matter. In the case of a non-accreting black hole then *dm/dt* = 0, and we recover the simple harmonic oscillator solution for the motion. Equation (20.1) is similar to the equation of motion when Stokes' drag term is included; in this case, however, the motion is not dependent upon the square of the velocity — which was the situation considered in Equation (17.4). The specific solution to Equation (20.1) is presented in the Appendix and it has the form:

$$r(t) = R \exp\left(-\frac{1}{2}\beta\omega t\right)\cos\left(\frac{\omega t}{2}\sqrt{4 - \beta^2}\right) \tag{20.2}$$

where R is the radius of the Earth and $0 \leq \beta \leq 2$. Equation (20.2) indicates that the black hole will oscillate back and forth across the Earth's interior (described by the cosine term) with steadily decreasing amplitude (described by the exponential term). The time-decreasing exponential decay term, in fact, dictates that eventually the black hole will come to rest at the Earth's center. The oscillation period of an accreting black hole will be longer than that of a simple harmonic oscillator with $T_{BH} = (4\pi/\omega)$ $(4 - \beta^2)^{-\frac{1}{2}} = (T_{shm})(1 - \beta^2/4)^{-\frac{1}{2}}$, and the settling time, when $r_{max} < R/1000$, will be after a time $t_{set} \approx 14/\beta\omega$ (see the Appendix for details). The rate at which the oscillating black hole accretes mass (under the conditions being considered here) will be $m(t) = m(0)\exp(\beta\omega t)$, where $m(0)$ is the initial mass. Figure 20.1 shows the time evolution of the motion for an accreting black hole moving through the Earth with a settling time of 7 days. This example takes $\omega = (g/R)^{\frac{1}{2}} = 1.240 \times 10^{-3}$ s^{-1} and $\beta = 0.019$, giving an e-folding time, in which the radius of oscillation decreases by a factor of $e = 2.71828...$, as $t_{ef} = 2/\beta\omega = 23.58$ hours; the oscillation period will be $T_{BH} = 84.46$ minutes. During the course of the motion the mass of the black hole will increase by a factor of $e^{14} = 1.2 \times 10^6$. In this example we have set the settling time to an arbitrary 7-day time interval, but a physically more realistic approach would be to consider what the likely accretion rate of the black hole might be, and from there deduce the appropriate value of β and the settling time [43]. Once the black hole has settled at the center of the Earth, however, it could continue to slowly, very slowly, accrete matter from the core, ultimately realizing Niven's imagined scenario for Mars in The Hole Man.

Chapter 21

The Elephant in the Room

With four parameters I can fit an elephant,
and with five I can make him waggle his trunk.

John von Neumann (attributed quotation).[1]

Up to this point I have carefully avoided any detailed discussion of the long and contentious history of *inventing,* via dazzling leaps of imagination, tortured logic, shear ignorance and outright fantasy, the Earth's interior. Such abstractions, beyond any reasonable reality, were commonly pronounced throughout the 19th century — a time of otherwise great conservatism in scientific matters. The hollow Earth; the Earth composed of concentric spheres; the Earth as the inside surface of a spherical shell; the Earth with a central Sun; the Earth with two central suns; all such malformations have been pronounced, but none have survived any detailed, even partial, exploration. Indeed, such flights of fantasy are typically conjectured upon the basis of some divine inspiration, and/or upon a less-than-credible misunderstanding of basic science and geology. Such is human nature, however, and while one is prepared to accept, in principle, the mantra that "blessed are the cracked for they let in the light," such ridiculous ideas as the hollow and/or the flat Earth, and the people that propagate them, do, in reality, immense harm to society.

The world is complicated enough without such buffoonery, and scientists do themselves no favors by silently and/or simply dismissing

---

[1] The profile of an elephant can, in fact, be generated with just four parameters — provided that they are allowed to be complex numbers. See www.johndcook.com/blog/2011/06/21/how-to-fit-an-elephant/.

such ideas as charmingly wacky. For wacky they are. The *alternative thinking* and shear chutzpa of such alternative *prophets* as John Cleves Symmes (see Figure 21.1), Cyrus Teed (a.k.a. Koresh [Hebrew for Cyrus] — see Figure 21.2), Marshall B. Gardner (recall Figure 9.2) and many others are not something that humanity should be especially proud of. Yes, there is the concept of free speech, but there is also delusional absurdity, and the latter is predicated on madness. Certainly, scientists have a moral duty to inform the general public, which typically finances their research institutions, but society also has the obligation to pay

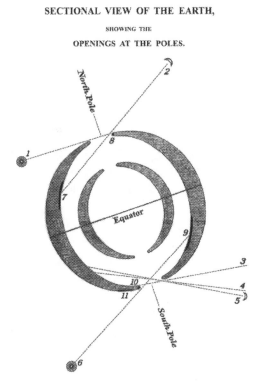

SECTIONAL VIEW OF THE EARTH,

SHOWING THE

OPENINGS AT THE POLES.

Figure 21.1. Frontispiece from *Symzonia: Voyage of Discovery* (1820) by Captain Adam Seaborn (attributed to be by J. C. Symmes). The diagram depicts the polar openings and shows an inner shell. Various features are numbered: 11 indicates the location of "Seaborn's land"; 10 indicates "Token Island"; 9 shows "Symzonia"; 8 is the "Place of exile"; and 7 is the "supposed place of Belzubia". The other numbers correspond to sight lines for the Sun (1 & 6) and Moon (2, 3, 4 & 5).

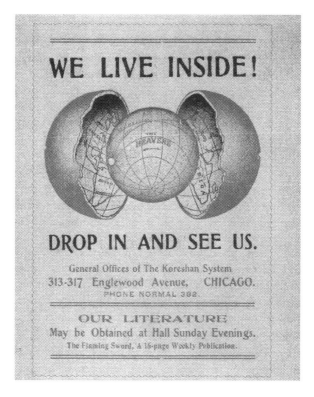

Figure 21.2. Advertisement for "Koreshan System" literature (Teed adopted the name Koresh in 1869). The picture suggests that we live on the inside surface of a sphere that in turn surrounds an inner sphere containing the Sun, planets and stars.

attention to what it is that the scientific research reveals. Society in general, and the individual specifically, does not have the unfettered right, nor the privilege, to simply pick, choose, ignore, twist and/or manipulate scientifically derived facts to their own ends and then expect to go uncriticized. Sadly, however, we continue to live, for so it seems, in a world described by philosopher Bertrand Russell in which "the average man would rather face death or torture than think." It is, in these modern times, the lack of critical reasoning that has resulted in the internet being clogged with conspiracy theories [1] and that, amongst many other important issues, global warming is denied.

With the above being said, it is appropriate that some justification be presented for calling the hollow-Earth theories as envisioned, for example, by John Symmes as mere fantasy [45]. For indeed, in spite of apparent parallels, they are very different to the Earth-tunnel thought experiment. The key reason for this dichotomy is in the use of the term "thought experiment." The Earth-tunnel thought experiment has resulted in the growth and development of scientific knowledge; in complete contrast the hollow-Earth theories are predicated on ignorance. Space and science-fiction artist David Hardy provides us with an analog of the situation. Indeed, Hardy has written that "artists have an advantage over mere technology, for they can travel where machines cannot — into the past, into the future, and faster than light" [46]. For us, it is certainly the case that thought experiments do have the advantage over technology (mere or otherwise), but they do not have the advantage of going anywhere that the human imagination might take them. The thought experiment can only operate within the realm, and within the strict constraints, of known physical principles. The hollow-Earth theorists, if we are being generous, fall foul of the physical constraints and take their thoughts to any location where their fancy pleases. Albert Einstein put it another way: "the difference between stupidity and ignorance is that genius has its limits."

Formal definitions are always complicated, since words are inherently slippery and they can carry different meanings, at different times, to different people and in different contexts. For all this, however, the OED offers the following useful definitions:

> *Thought*: "process, power, of thinking; faculty of reason, sober reflection; consideration."
> *Experiment*: "test, trial, procedure adopted on chance of its succeeding or for testing hypothesis."

The concatenation of these two definitions provides an entirely reasonable description of what a thought experiment is. Yeates (University of New South Wales, Australia) [47] further elucidates, in more refined philosophical tones: "A thought experiment is a device with which one performs an intentional, structured process of intellectual deliberation in order to speculate, within a specifiable problem domain, about potential consequences (or antecedents) for a designated antecedent (or consequent)." This more formal definition begins to both constrain and pin down exactly what can and cannot be done with respect to a thought experiment. Yes, a thought experiment can be, and indeed is a flight of fantasy, but it must be an especially constrained flight of fantasy that does not contradict any of the known laws of physics that have been shown to successfully describe the workings of the real world. The historical success of thought experiments (as the proceeding chapters indicate) is indeed remarkable, and while still a topic of continued discussion amongst philosophers, the successes appear to indicate that a purely cognitive investigation can be used to tease out and quantify the workings of an otherwise hidden objective reality. Importantly, however, thought experiments are nothing like the dreams of Peter Ibbetson and Mary, the Duchess of Towers, described in the highly successful book (and later play) *Peter Ibbetson* by George Du Maurier (first published in 1891). Du Maurier tapped into the apparent Victorian need for a belief in the supernatural and his story revolves around a series of dreams which connect Peter and Mary. These dreams enable the protagonists to interact and travel through time together — the mind effectively transcending the reality of their spatial and temporal separation. There is a nomological reality about thought experiments, that is, when all is said and done, they must be realizable at least in some abstract, if not definite, sense.

Thought experiments can stretch and push the limits of the possible, and they can illuminate concepts otherwise hidden in shadow. They are the tools of mind exploration, and thought experiments can take physics into those far, deep and many-dimensioned realms completely closed off to direct experimentation. And yet, for all this power, thought experiments can also be exceptionally simple in their design. Kepler's thought experiment, outlined in his posthumous *Somnium* (published in 1634), is predicated on the Copernican model and he takes the reader on an imagined journey to the Moon and from there to a vista containing a revolving Earth. Galileo asked the readers of his *Dialogue Concerning the Two Chief World Systems* (published in 1632) to consider the motion of objects confined within the interior of a ship; first in a stationary state and then moving — introducing in this manner the idea of Galilean relativity. Newton, as we have seen, asked the readers of his *Principia* (published 1687) to consider numerous thought experiments, thereby taking their minds and understanding to wholly new domains of physical reality.

Some thought experiments have deep philosophical undertones and begin to explore the potential limits of knowledge. Pierre-Simon Laplace is remembered in this sense for his 1814 invocation of a demon that knows (at some specific instant) the exact location and momentum of every atom in the universe (no small task, of course). If this were the case, then Laplace's demon, in principle, would be able to calculate through the laws of classical mechanics every past and future position of every object in the universe. The demon conjured up by Laplace would therefore be able to calculate exactly how the past brought us to the present and how in turn the present will take us to a certain future. For the demon the universe is entirely deterministic. Under this imposition, the fact that we observe events that appear to take place at random is entirely a consequence of our ignorance about the physical laws that govern the universe — laws that are only known to the demon. That such an all-knowing demon cannot exist relies on the development of ideas concerning quantum mechanics (specifically Heisenberg's uncertainty principle), chaos theory and the inherent uncertainty that will exist in any measurements that can conceivably be made — that is, it is simply not possible to know, with infinite precision, the exact position and momentum of even a single atom, let alone all of the atoms in the universe.

In similar vein to Laplace, James Clerk Maxwell, in his *Theory of Heat* (published in 1872), while a master of mathematical analysis, asked his readers to consider a thought experiment concerning the statistical nature of the second law of thermodynamics. Indeed, Maxwell invoked another demon to complement that conjured up by Laplace; an imagined entity in this case, however, that selectively separates out fast and slow moving gas particles into two chambers. In principle this demon could engineer a system in which the entropy actually decreases — violating the second law of thermodynamics (which requires that entropy can never decrease). There is still no fully accepted and/or philosophically water-tight answer to Maxwell's thought-invoked demon, but the standard answer for its non-existence lies within the fact that even a demon must do work and expend energy in order to know something about the gas particles that it is attempting to separate out, and accordingly it is the increase in entropy of the demon, and the fact that the demon and gas are interacting, that saves the second law from ever being violated.

Erwin Schrödinger, famous for his many contributions to physics, is particularly well known for his thought experiment concerning a hapless cat that may or may not be alive within a closed box. Developed in response to the Copenhagen interpretation of quantum mechanics, Schrödinger's cat thought experiment cuts to the very core of quantum physics, asking when and how a superposition of quantum states becomes a uniquely observable classical state.

Perhaps the grandmaster of the thought experiment, however, is Albert Einstein: indeed, his great mind has possibly generated more subtle thought experiments than anyone else since Isaac Newton. Special relativity, for example, is predicated upon a thought experiment in which an observer attempts to chase a beam of light; his general relativity is further predicated upon another thought experiment in which an observer is imagined to be trapped within a closed elevator — the point being that such an observer will not be able to distinguish between the effects of gravity and that of accelerated motion.

Other thought experiments are futuristic, in principle entirely realizable, but well beyond present human technology. Freeman Dyson's notion that an advanced civilization might surround its parent star with energy-tapping islands (recall Chapter 9) is a thought experiment that has manifestly resulted in the broadening of concepts with respect to how advanced civilizations might be detectable within our galaxy, but no such structure has ever been seen.

Hollow and flat-Earth proponents, both in the past and in the present day, are not performing thought experiments. Indeed, they are not performing any kind of experiment or describing any logically derived system of ideas based upon observations or physical theory. They are either peddling unfettered dogma or promoting misinformed delusion. In the sense of being falsifiable, a stringent condition imposed by Karl Popper upon all scientific theories, the hollow-Earth concept is in principle testable, but, of course, the advocates of such theories simply don't (at least historically) accept any experimental result that runs contrary to their dogmatic dictates. It is the inherent failure of such protagonists (as well as the equally misguided conspiracy theorists [1]) that they are incapable of seeing beyond their own opinion. This, of course, is just delusion, and it is a sad delusion at that, since the universe is a much more interesting and exciting place than their blinkered dictates allow for.

Science is not a faith-built structure, nor is it delivered according to any divine revelation; rather, and most importantly, it is a progressively derived and independently time-tested series of arguments that have a basis of certainty within a clearly defined framework of observations and measurement. It is these very specific conditions that separate out, for example, Edmund Halley's hollow-Earth model as a scientific theory, from those of say John Symmes or Cyrus Teed, which are delivered according to divine fiat (Figure 21.3). Halley's theory, as described in Chapter 15, was scientific in the sense that it was formulated upon a carefully derived set of observations, and the internal shells had a specific purpose within the framework of explaining Earth's magnetic field configuration. That Halley's model is no longer used today immediately informs us that we now have a much better model, and a much better understanding of both the interior of the Earth and the way in which the geomagnetic field is generated — which is not to say that we know everything in detail.

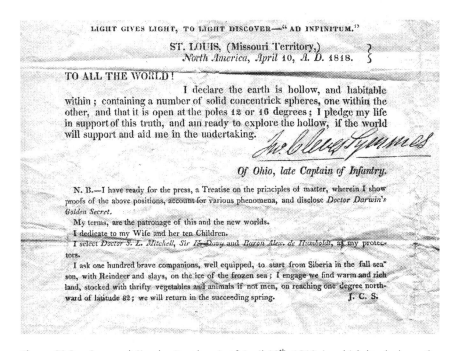

Figure 21.3. Symmes' Circular Number 1, of April 10th, 1818, in which he declares the Earth to be hollow and in which he asks for help in mounting an expedition to its interior.

Science does not allow for the luxury of cheating and/or simply making up answers to explain the gaps that exist in our knowledge about the universe.

Physicist Richard Feynman has saliently written that "the first principle [of science and life] is that you must not fool yourself and you are the easiest person to fool." Within this framework Samuel Rowbotham provides us with a telling historical example of just how a flat-Earth proponent can fool both themselves and, in the process, many others. The case in point is that of the Bedford Level canal experiments conducted in the early 1900s. Indeed, it was Rowbotham (who adopted the *nom de plume* Parallax) that initiated a public challenge, arguing that when he had carefully observed a boat traveling along a six-mile (9.65 km) stretch of the canal there was no variation in the visibility of its hull and that its five-foot high mast was always visible. The boat did not disappear "hull down" as was required by a curving Earth's surface, and accordingly Rowbotham declared that the canal, as well as the Earth, must be absolutely flat (recall Chapter 8 and Figure 8.2). Taking the Earth to be a sphere of radius $R = 6{,}371$ km, the curvature of its surface would dictate a drop in height h given by the equation $\cos \varphi = R/(R + h)$, where $\varphi = (9.65/6371)(180/\pi)$ degrees. This implies that $h = 7.3$ meters which is well in excess of the five-foot (1.52 m) mast on the boat. With his *observation* in hand, Rowbotham set about traveling the length and breadth of England promoting the revelation that the Earth was flat and that science had got it all wrong. He even produced a pamphlet (later expanded into a full-length book) on his observations under the title *Zetetic Astronomy* (1849).[2] Due to strong public interest (not to mention the growing pressure from his more vocal skeptics) Rowbotham eventually agreed to conduct another flat-Earth demonstration at Plymouth Hoe in 1864. The target in this case was the Eddystone lighthouse which is located some 14 miles (22.5 km) offshore. If the Earth was flat then the entire lighthouse should be visible through a telescope located on the Hoe. The fact that the lighthouse was clearly not fully visible to all those who looked through the telescope did not phase Rowbotham in the slightest, and he continued to simply claim the opposite observation.

---

[2] Rowbotham coined the word zetetic from the Greek *zeteo*, which means to search or examine; that is, truth seeking.

One might expect that at this stage Rowbotham and his flat-Earth conceit would fall by the wayside, but far from it, and remarkably other flat-Earth supporters came to the fore and this time money was put on the table. In 1870 John Hampden announced a £500 wager that he could exactly replicate Rowbotham's experiment on the Bedford Level canal. The wager was to be matched by any challenger supporting the opposing view, and the observations would be made under the scrutiny of two referees. Such wagers, of course, have absolutely no scientific weight and are only engineered to provoke public interest. Hampden's challenge was eventually taken up by Alfred Russell Wallace,[3] who with an experienced background in surveying techniques easily showed that Rowbotham's original claim was entirely untrue, and that the heights of various marker poles, placed along the canal surface, did indeed show a systematic variation in their appearance in accord with expectation for a curving Earth's surface. The referees (after some acrimonious squabbling) awarded Wallace the victor's prize even though Hampden refused to accept the demonstration as being valid. Indeed, Hampden sued for his money back and even published a pamphlet arguing that Wallace had cheated and blatantly fixed his results. Eventually, after a number of court challenges, Hampden was jailed for libel. That Wallace had indeed provided a correct demonstration of the curvature of the Earth was vindicated later through a series of refined measurements by Henry Oldham in 1901 [48].

It is interesting to note, and maybe it was no accident, that shortly after the Hampden–Wallace debacle was concluded, school master and theologian Edwin Abbott Abbott brought out his now famous and delightful, *Flatland: A Romance of Many Dimensions* (Seeley & Co. of London, 1884). This book (Figure 21.4) explores the idea of a two-dimensional universe, *Flatland*, in an entirely imaginative, but logically consistent manner. It is a controlled flight of fantasy, with deep-running undertones that reflect upon the hierarchical nature of Victorian society and the limits of human understanding. The text also explores aspects of super-dimensional space along with issues pertaining to human credulity and the belief in supernatural phenomenon.

---

[3] This is the same Alfred Russell Wallace who along with Charles Darwin pioneered the theory of evolution by means of natural selection.

Figure 21.4. From the front cover of Edwin Abbott Abbott's *Flatland*. In the *cloud* surrounding flatland are labels pertaining to multi-dimensional spaces. Abbott's dedication makes specific note to the "citizens of that Celestial Realm" that may aspire to "the secrets of FOUR FIVE OR EVEN SIX Dimensions".

For all that Oldham had usefully advanced public understanding with respect to the results of the Bedford Level experiment, Lady Elizabeth Blount managed, in 1904, to turn that understanding once again into controversy. Specifically, Lady Blount hired a photographer to repeat Rowbotham's experiment, and, of course, the claim was made that the new observations vindicated the flat-Earth interpretation. Readers of the *English Mechanic and World of Science* [2][4] first heard of this new observation in the August 19th issue for 1904, and the debate between its various readers, both pro and con, was to rage for more than a year. Indeed, on September 16th, 1904, Lady Blount was to write within the *Mechanic's* pages, "With the aid of the latest discoveries and improvements in the art of photography, the Earth's unglobularity is proved." The following

---

[4] By 1904 the magazine had changed *Mechanics* to *Mechanic*.

issue of the *Mechanic* saw seven letters of denouncement — and so the debate continued. The *Mechanic* eventually published Lady Blount's photograph of the Bedford Level canal in its October 28th, 1904 issue, but this only intensify the letters of protestation from opponents on both sides of the argument. Lady Blount was indefatigable, and went on to become one of the founders of the Universal Zetetic Society, and an early editor of its magazine, *The Earth Not a Globe Review*. Nearly a century after its formation the Universal Zetetic Society became the present-day Flat Earth Society in 1956. Why, we must now ask, did Rowbotham, Hampden and Blount, and many others, feel so strongly about establishing the flat-Earth proposition? And, the answer lies in the objectives outlined in the formation documents of the Universal Zetetic Society — "the propagation of knowledge related to Natural Cosmogony in confirmation of the Holy Scriptures, based upon practical scientific investigation." In other words, the flat-Earth advocates, at least in the early 20th century (if not in the present-day) had an agenda, and the agenda was based upon a literal (and uncritical) reading of the Bible, which supposedly requires that the Earth be flat. With this as a foundation principle there is no scientific debate to be had, and indeed, no scientific statements under its banner can be made either. The flat-Earth defenders are no different to the present-day creation, or intelligent design advocates who delude themselves and fail to see the delusion. Their use of scientific terminology and apparent scientific methodology is nothing more than a disingenuous ruse to promote and attempt to justify unchangeable opinion.

The flat-Earth proposition is typically predicated upon an attempt to justify a biblical story, and it is not a scientific or logically reasoned proposal with any merit in the modern era. This is not to say, once again, that scientists working within the framework of scientific discussion and experimentation have not made bold, and even ridiculous-sounding claims. While it is entirely inappropriate to judge the collected scientific works of Scottish mathematician and physicist Sir John Leslie (1766–1832) in terms of just one three-page "end note" discussion, it is the case that he made the argument in 1829 that the Earth must have a hollow structure. Writing in his *Elements of Natural Philosophy* (published by Oliver and Boyd, London), Leslie added in relation to his lengthy discussion concerning the compression of matter a few "end note" points with respect to the structure of the Earth. At that time Leslie and others were just beginning to understand the effects of compression upon matter. Writing at a time long before either atomic structure and electromagnetic radiation were understood, Leslie worried about the possibility that matter might be compressed to higher and higher densities, and he specifically worried about the possibility that matter inside of the Earth might be compacted to near infinite density. In an attempt to avoid this possibility coming about, he suggested in his *end note* discussion that "our planet must have a very widely cavernous structure, and that we tread on a crust or shell whose thickness bears but a very small proportion to the diameter of its sphere." He then goes on to argue that the only substance that has both the diffusive and non-compressible properties that could exist under the conditions he envisioned to operate within the Earth's interior was light itself. He concludes, "we are thus led, by a train of close induction, to the most important and striking conclusion. The great central cavity is not that dark and dreary abyss which the fancy of poets has pictured. On the contrary, this spacious internal vault must contain the purest ethereal essence, LIGHT in its most concentrated state, shining with intense refulgence and overpowering splendor."

Leslie's idea is far-fetched, but is not inconsistent with the then understood physics of light and matter. The fact that we now no longer have the same concerns as Leslie, and indeed, no longer give his suggestion any credit, is that from our time-advanced perspective we have a much better understanding of how atoms, molecules, solids and light work. Science is provisional and science is self-correcting, indeed, it is the exact opposite of creation and/or intelligent design concepts. It is the very provisional and self-correcting nature of science that makes it such a successful, powerful and enlightening enterprise, and indeed, it paves a sturdy and dependable path towards an ever better understanding of reality — which is not to say that any complete understanding will be arrived at easily. Indeed, the only reason that we know about Leslie's ideas of a hollow Earth to this day is through a peripheral mention of it in Jules Verne's *Journey to the Center of the Earth* (1864, and recall Chapter 9). Verne writes in Chapter 30 of his *Journey*, "then I remembered the theory of an English Captain, who likened the Earth to a vast hollow sphere, in the interior of which the air became luminous because of the vast pressure of weight upon it, while two stars Pluto and Prosperine, rolled within upon the circuit of their mysterious orbits." With Verne's account we have a mixture of literary fusion and fiction, set as it is between the hollow-Earth ideas of Edmund Halley (who was a Captain — recall Chapter 15), Leslie's compression and inner light suggestion, and an entirely made-up pair of inner suns mentioned by neither Halley nor Leslie [49]. In ensuing years, while the scientific community was able to move beyond the ideas of Halley and Leslie, introducing more refined models and physical concepts as they became available, the fantastic writings of Verne remained fixed and stagnant (as must be the case with literature, but not so with science). It is Verne's made-up story, however, that has continued to catch the public imagination.

While Professor Leslie reasoned his way, with the application of then known physics, to a hollow, light-filled Earth, a later work by William F. Lyon draws the same conclusions but does so by invoking confusion and delusion. The work by Lyon, *The Hollow Globe: or the World's Agitator and Reconciler. A treatise on the physical conformation of the Earth* (Religio-Philosophical Publishing House, Chicago) appeared in 1871, and it is a work of pseudo-science (Figure 21.5). In the introduction, Lyon cuts to the core of his thesis, "this globe [the Earth] is constructed in the form of a hollow sphere, with a shell some thirty to forty miles in thickness, and that the interior surface which is a beautiful world in a more highly developed condition than the exterior, is accessible by a circuitous and spirally formed aperture that may be found in the unexplored open Polar Sea" [recall Chapter 9 and especially James De Mille's *A Strange Manuscript Found in a Copper Cylinder*). Indeed, the text reads like some adventure/mystery novel, with the main details of the story being conveyed to Lyon by a mysterious stranger who chanced to visit his office one day. As a piece of imaginative literature Lyon's text makes for amusing, if not interesting reading since it twists and turns between what are established facts and observations and unadulterated make-believe and delusion. Lyon's book and basic thesis is no more than the promulgation of a utopian dream.

A similar such utopian text was produced by Marshall B. Gardiner, *A Journey to Earth's Interior* (published by the author at Aurora, Illinois) in 1920. Once again, however, Gardiner uses pseudo-scientific language to

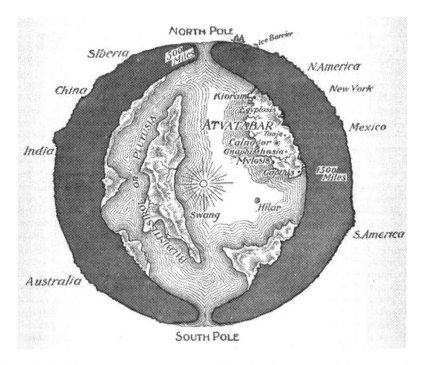

Figure 21.5. There were no actual diagrams in Lyon's text, but several later reproduction issues have used a cover illustration from William Bradshaw's 1892 novel, *The Goddess of Atvatabar: being the history of the discovery of the interior world, and conquest of Atvatabar*. In his essentially utopian story, Bradshaw largely evoked a Symmesian-like inner-Earth topology (recall Figure 21.1). There is a central star, called Swang, a moon (or "wandering sphere") called Hilar, and access openings to the outer surface in the north and south polar seas.

develop his ideas. Early on in his text Gardiner writes, "we claim that the Earth is a hollow body with an immense opening at each polar axis — an opening about fourteen hundred miles in diameter and that there is in the interior of the Earth a Sun which warms it and gives it light." The thickness of the Earth's outer shell is estimated as being 800 miles, while the inner Sun has a diameter of possibly 600 miles. To justify this structure, Gardiner argues that the solar nebula theory requires modification, and that, in fact, all planets and their attendant moons form as hollow structures. Interestingly, Gardiner attacks John Symmes's earlier 1818 text on the Earth's hollow structure, arguing that "Symmes has no coherent theory," and that his arguments are based upon unsupported "facts" [50]. In contrast, Gardiner suggests, invoking El Dorado-like imagery, "we have opened the road to a new world in our theory and it must be a world of inconceivable richness." Indeed, Gardiner set out to protect his ideas by applying for a patent on his new globe (recall Figure 9.2) — a globe that he envisioned would be needed in every school in the world. Gardiner concludes his text with the plea, "for economic reasons, then, as well as for the advancement of science and the glory of discovery, it is of the utmost importance that the interior of the Earth should be explored." Gardiner's sentiment is perhaps correct in only one sense, and that is that the Earth should be explored for the advancement of science.

Chapter 22

First and Last Thoughts

Roads go ever ever on,
Over rock and under tree,
By caves when never sun has shone,
By steams that never find the sea

J. R. R. Tolkien

The story of the Earth-crossing tunnel is as long as it is deep and as timely as it is timeless. Indeed, the narrative of falling through the Earth is as old as the hills and as recent as the present day. The Earth-tunnel experiment has straddled the bounds of history, and it has played a fundamental role in the historical development of ideas relating to dynamics and the basic physics of how objects fall and move. Additionally, the Earth-tunnel question, in spite of its inherent impossibility, qualifies as an important thought experiment, and reflecting upon its implications has steered numerous historical figures, including such *illuminati* as Thomas Bradwardine, Galileo Galilei, Isaac Newton, Robert Hooke, Edmund Halley, and Leonhard Euler, towards a better understanding of the Earth and the mathematics of accelerated motion. Today, however, the Earth-crossing tunnel is seemingly reduced in its stature, and it tends to be known for its introductory capacity as a question on oscillatory motion — indeed, it is founded upon the prize-winning question first posed by Nevil Maskelyne (Terricola), the 3rd Astronomer Royal, in 1781. This contemporary devaluation of the Earth-tunnel problem is indeed a tragedy, since it has such a rich and vibrant history: a history that includes innovations in mathematics, the development of new physical theory, new engineering principles, philosophy, and the geological appreciation of Earth's inner structure. It also has a deep cultural history, both with respect to mythology, human experience, literature and cinematography. The Earth-tunneling problem has much more to reveal to the student of science than its present pedagogic application, good though that application is, seemingly allows for.

Not just on Earth, however, Cymro's problem (stemming from Terricola's question) has applications in the greater solar system and space physics. Indeed, one space-based application of the Earth-tunneling experiment, which might one day be made into a real experiment, is that outlined by David Berman and Robert L. Forward[1] in 1968. In the suggested experiment the Earth (as such) is cut down in size to a uniform-density sphere of about half a meter or so in diameter, and through this sphere a hole, now in an entirely practical manner, is drilled. The sphere is then taken into space and placed in a spacecraft or space-station laboratory thereby setting in a free-fall, low-Earth orbit. The Earth-tunneling experiment can now proceed on a miniature scale. The key aspect of the experiment is to measure the oscillation time T of a small bead placed within the borehole cutting across the mini-Earth interior. With this timing measurement secured the universal gravitational constant G can be determined as:

$$G = \frac{3}{4\pi} \frac{\omega^2}{\rho\Delta} \tag{22.1}$$

where $\omega = 2\pi/T$ is the measured angular frequency, ρ is the density of the sphere and the Δ term is explained below. With this experiment an estimate of the notoriously difficult to measure gravitational constant G can be made, and the uncertainty in the measurement could, at least in principle, be made very small since it is a timing measurement only and these can be made to a very high order of accuracy. Indeed, the accuracy of this experiment could (in principle) exceed that which is obtainable in terrestrial laboratories. As with all experiments, however, there are always

---

[1] This is the same Robert L. Forward who trained as an aeronautical engineer, but became famous for his many science fiction writings and contributions towards the development of interstellar travel.

compromises and construction issues that need to be resolved. While not impossible, it is very difficult to produce truly uniform density solids, and it is likewise difficult, but not impossible, to make very accurate spherical profiles. Indeed, in their 1968 proposal, Berman and Forward noted that, at that time, the most difficult part of the experiment would be to manufacture a truly uniform density sphere. In the modern era this is far less of a problem. Deviations of the mini-Earth profile from sphericity can be gauged by considering the gravitational attraction of a non-rotating oblate spheroid [51].[2] In this manner if the spheroid has an equatorial radius of a, and a polar radius $b < a$, then the gravitational force acting on the tunnel-crossing particle will be $F(r) = [(4\pi/3)G\rho\Delta]r$, where $\Delta = (1 - f/5)$ in the polar-crossing case, and $\Delta = (1 + 3f/5)$ in the equator-crossing case, with $f = 1 - (b/a)$. In the perfect sphere case when $f = 0$, $\Delta = 1$. The Δ correction term enters the final calculation for G in a linear fashion — see Equation (22.1) — and the uncertainty is entirely dependent upon the f-term being as close to zero as possible. A further correction that would need to be assessed in any high-precision measurement would be that relating to the effect of the tunnel itself, and how the removal of material from the channel changes the overall gravitational field experienced by the test particle. Such a calculation is non-trivial, and Paul Worden and Francis Everitt have more recently suggested that perhaps a long hollow cylinder, rather than a sphere, might be a more appropriate substrate to use [52]. The equation of motion for a particle dropped into a hollow cylinder is far more complicated than that for a sphere, but it is nonetheless expressible analytically.

As a thought experiment the Earth-tunneling problem has provided a rich bounty of fundamental demonstrations and insights. As a practical engineering exercise, however, the Earth-tunneling problem has been, and will likely remain so, a complete and utter non-starter. Even if the mechanical excavation devices could be constructed there would be no practical (or fiscally reasonable) reason for building an Earth-crossing tunnel or a gravity train system — there are simply many other cheaper ways of getting about upon the Earth's surface. The situation elsewhere

---

[2] This topic has an interesting parallel to Terricola's Prize Question of 1801 (see the Appendix).

within the solar system, however, may yet, in the future, prove much more practical and financially viable.

There have been two Hollywood movie adaptations to Philip K. Dick's 1966 short science-fiction story *We Can Remember It for You Wholesale* [53]. Both films were called *Total Recall*: the first from 1990 starred Arnold Schwarzenegger and was directed by Paul Verhoeven. The second film version came out in 2012, starred Colin Farrell and was directed by Len Wiseman. Neither film was a blockbuster hit, but each introduced a dramatic action-driven storyline in which reality, unreality, true memories and implanted false memories were set against a backdrop of a future dystopian world. The first *Total Recall* movie was largely set on Mars, while the second, being more in sympathy with Dick's original story, was set on a post-biochemical-war-ravaged Earth. In the latter version, only two regions of the Earth's surface remained habitable, that of the United Federation of Britain and that of The Colony (Australia). The second *Total Recall* movie, and Dick's original narrative, employed an Earth-crossing tunnel (called *The Fall*) which enabled people to rapidly commute between the two habitable realms of the Earth. Ultimately, however, as the story unfolds, *The Fall* is destroyed and presumably all direct communication between the hemispheres is lost. This would indeed be a disaster in the context of the world setting invoked in Dick's narrative, but one assumes that ultimately some form of appropriate land, air and sea-going transportation would be developed and employed to enable direct contact again. This development of alternative transport systems, however, might not be so easy on another planet. Bodies such as the Moon, Mercury and even the larger asteroids could in principle be circumnavigated by land transport; these worlds having no atmosphere to support wing-powered transport (rocket-powered ships might presumably be useable). Mars and Venus, even in a non-terraformed state, will allow for land and aerial transportation. In a terraformed state, both of these worlds might additionally allow for ocean-going transportation [54].

The tunnel-transit times for various solar system bodies are presented in Table 22.1, and while Mars is about half the size of the Earth, the transit time across its interior is some 8 minutes longer than that of a terrestrial gravity train (taking about 50 minutes). The crossing time for

Table 22.1. Tunnel-transit times for a selection of solar system bodies. The columns provide an identification of the solar system body, followed by its corresponding size and surface gravity. The last two columns give the crossing times in minutes and the maximum velocity in kilometers per second attained at the object's center.

Object	R (km)	g (m/s²)	T_{cross} (min.)	V_{max} (km/s)
Sun	696,265	291.268	25.77	450.33
Mercury	2440	3.727	42.37	3.02
Earth	6378	9.807	42.23	7.91
Mars	3396	3.727	49.98	3.56
Ceres	480	0.275	69.18	0.36
Jupiter	71,492	24.812	88.88	42.12
Saturn	60,268	10.395	126.08	25.03
Pluto	1195	0.657	70.62	0.89

Earth's Moon is 53.5 minutes (see the Appendix discussion concerning *The Scientific Adventures of Baron Munchausen* by Hugo Gernsback). The crossing time for Ceres, the largest of the asteroids located between Mars and Jupiter (at some 930 km across) is just under 70 minutes. The crossing time for Pluto is also about 70 minutes as well. These latter worlds, having no atmosphere, come closest to the ideal tunnel-crossing scenario, and they may yet, in the future, be the first worlds in which a diametrical tunnel could prove advantageous and be feasible as an engineering exercise.

Where future transit-tunnel systems may well find useful application, as a form of communication and rapid transportation, is in the field of asteroid mining. Even in the present era, the concept of mining minerals from small solar system bodies is being actively developed by a number of venture companies. Remarkably, for so it appears, it may soon be the case that mineral extraction from asteroids will be more cost effective than mining material from the deep Earth [55]. Almost as a byproduct of the space mining process, therefore, one can envision a gravity-train system being developed within the interior of an asteroid. Such a tunnel system could then, in principle, be used to transport ore across the asteroid for no additional cost. Additionally, in the perhaps not-so-distant future, the self-burying idea (recall Chapter 19) and/or robotic drilling

Figure 22.1. An artist's impression of a robotic probe exploring the under-surface ocean of Jupiter's moon Europa. The self-sinking capsule (recall Chapter 19) used to penetrate the icy crust of Europa can be seen to the upper left in the image. Image courtesy of NASA.

might also be used to explore the interior regions of the ice moons Europa and Encelladus. These two moons being of particular interest since they both appear to support sub-surface liquid-brine reservoirs where bacterial (extremophile) life may have evolved independently of life on Earth (Figure 22.1).

Some experience of deep-core drilling through compacted layers of ice has already being gained on Earth in the frozen desert of Antarctica. Situated some 1,300 km from the geographic South Pole at the center of the East Antarctic Ice Sheet, Vostok Station was established in 1957, during the International Geophysics Year, and has been described as the most remote place on Earth. Indeed, it was at this Soviet Antarctic research station that the coldest temperature ever recorded on Earth was measured — a crisp −89.2°C on July 21st, 1983. Not only does Vostok Station hold this chilling record, it also holds the record, obtained in 2012, for the deepest borehole drilled through ice — borehole #5G-2

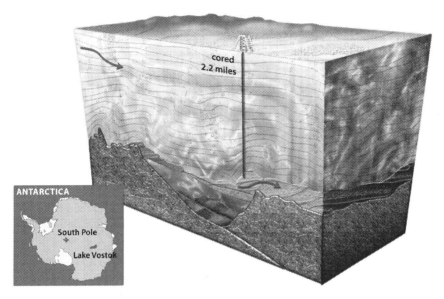

Figure 22.2. An artist's cross-section schematic and location map for Lake Vostok, Antarctica. The lake is covered by some 4 km of ice and has been isolated from surface sunlight for some 15 million years. Image courtesy of the National Science Foundation.

extending to a depth of 3,769 meters. Purely by chance Vostok Station is situated over the largest of the subglacial lakes: Lake Vostok. Being some 250-km long and 50-km wide, Lake Vostok is overlain by some 4 km of accumulated ice, and has been cut off from sunlight and the outside world for the past 15 million years (Figure 22.2). The lake is divided into two deep basins around a graben structure located within a rift valley that began to form some 35 million years ago. At such depths and pressure, the water in the lake is expected to be supersaturated with nitrogen and oxygen, and indeed, rival those conditions expected under the ice sheets of Jupiter's frozen moon Europa.

The very first water samples were collected from Lake Vostok in 2012 and the indications are that the lake, in spite of its lengthy isolation, supports a rich ecosystem. Indeed, the detailed analysis of the water samples revealed the presence of 3,507 unique nucleic acid sequences [56]. These sequences can mostly be identified with known bacteria and some indicate the presence of multicellular eukaryotes — there are additional sequences, however, that hint at new and presently unidentified bacterial life forms. The continued collection of pristine water samples from this ancient buried lake, situated at the literal end of the world, will no doubt yield more revelations and surprises, giving us a glimpse into the workings of new and hitherto unexplored worlds.

To the present day only the human imagination has traveled to the Earth's core. Indeed, Earth's interior is profoundly unknown and perhaps unknowable in terms of tactile exploration. Humanity's deepest mines and boreholes have hardly scratched the Earth's surface. Gravity trains and Earth tunnels are grand engineering ideas; they are pleasing and informative thought experiments, but it is difficult to imagine that they will ever be realized in tangible form. Will the Earth's interior ever be deep searched? Not, it would seem, anytime soon and not even in the foreseeable remote future. Indeed, our collective high-technology gaze is largely directed outwards, into space, rather than downwards towards the Earth's interior. Space may not, in spite of Captain Kirk's famous claim, be the "final frontier." Rather, for all indications point this way, the final frontier for exploration is literally going to be the ground beneath our very feet.

Appendix

Mathematical Details

I would earnestly warn you against trying to find out the reason for and explanation of everything [...][it] leads to nothing but disappointment and dissatisfaction, unsettling your mind and in the end making you miserable.

Queen Victoria (1883)

These appendix examples can be safely ignored by those readers, such as Queen Victoria, not wishing to see the mathematical details. They are, however, in spite of royal dismissal, instructive and worthy of consideration since they not only illustrate some of the great power of mathematics, they also reveal the limitations of the solutions at hand. The pioneering astrophysicist Arthur Eddington wrote in 1928 that "proof is the ideal before whom the pure mathematician tortures himself. In physics we are generally content to sacrifice before the lesser shrine of plausibility." The examples that follow are offerings to Eddington's "lesser shrine": they are classics, and while many were first posed over a century, or more, ago, they continue to find their way, in a meaningful sense, into both mathematics and physics textbooks of the present day.

A stone in a well

This classic problem has a certain amount of actual street-credibility, for indeed, one convenient way of gauging the depth of a well is to drop a stone into its opening and measure the time between the moment of release and the moment of hearing the splash. If we assume that the acceleration due to Earth's gravity is constant over the depth h of the well, and that the stone is let fall from rest (u = initial velocity = 0 m/s), then Galileo's equation of motion (see Chapter 12) provides the result that: $h = ut_F + \frac{1}{2} gt_F^2$, where t_F is the fall time for the stone. All well and good, but remember it is the time between release and hearing the sound of the splash that we are measuring. When the stone hits the water, the sound will propagate upwards at a speed c, the speed of sound in air, and it will do so over a distance h, so the sound travel time from the bottom of the well to the top is: $t_S = h/c$. What we, as the observer, measure is $t_F + t_S = t = (2h/g)^{\frac{1}{2}} + h/c$. To solve this final equation we can make the substitution $h = H^2$, to obtain a quadratic in H, such that: $H^2 + Hc(2/g)^{\frac{1}{2}} - ct = 0$, and this has the solution that $H = -(c/2)(2/g)^{\frac{1}{2}} [1 \pm (1 + 2gt/c)^{\frac{1}{2}}]$. At this stage we can substitute for the acceleration due to gravity at the Earth's surface: $g = 9.81$ m/s^2 and the speed of sound: $c = 343.2$ m/s (in dry air, at 20 degrees centigrade, at sea level). Accordingly, the well depth h can be determined for any given splash-sound time t — see Figure A.1. So, for example, if the splash is heard 2 seconds after releasing the stone, the well is about 19 meters deep. If the splash is heard 10 seconds after releasing the stone then the well is 390 meters deep. It will take some 17 seconds to hear the splash from a stone dropped into a well 1-km deep. The speed of the stone when it hits the well-water surface is given by the expression $v_{hit} = u + gt_F$.

The fall time of the stone and the travel time of the splash sound wave will be equal at a depth of $h = 2c^2/g \approx 24$ km, when the fall time is equal to 70 seconds, and $v_{hit} \approx 687$ m/s (a speed of Mach 2). With this final calculation, however, we have pushed our analysis much too far. At such great depths (even if such a well could be dug — and as we have seen it cannot) we are no longer safe in our assumption of constant gravity, and we can no longer ignore air resistance, which will drive the stone towards a constant terminal velocity, and this quantity will depend upon both the shape and the mass of the stone (see Chapter 17). Likewise, over such long well depths, the speed of sound will no longer be constant since it varies according to the square root of the ambient temperature — sound waves propagate more slowly (in an ideal gas) when the temperature is higher. Not only this, but it is highly unlikely that a sound wave could propagate multiple tens of kilometers inside of

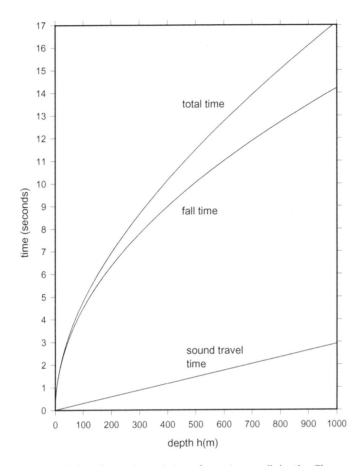

Figure A.1. Stone fall and sound travel times for various well depths. The upper curve shows the total time between dropping the stone and hearing the splash. The fall time corresponds to the time from release to hitting the water; the sound travel time corresponds to the sound passage time upwards to the well opening.

a borehole and still be audible to the human ear at the entrance. We have additionally ignored the spin of the Earth in our solution, and this, of course, will inevitably drive the stone towards a collision with the borehole wall, and any such collisions will alter the value of the fall velocity. With these latter issues we find the complex limit to what is otherwise a simple physical problem, and only a detailed numerical investigation will take us further.

Into the bottomless pit

The following problem, question 154 in fact, was proposed by the "Rev. Mr. T-H-" in Benjamin Martin's *Miscellaneous Correspondence* in October 1757: "Suppose a bullet fall down eternally in this manner, the first minute 20 miles, the second 19 miles, the third 18 1/20th miles, and so onward forever in the same geometrical progression, how far will it fall in all whole eternity?" There is no indication why the good reverend was interested in such a calculation, other than as a pure mathematical exercise, although one might speculate that he could have had an interest in the bottomless pit (reserved for the disembodied spirits of the pre-Adamite world) described in *Revelations 9:2*. The question is clearly contrived but it leads to an intriguing idea in its result — the idea being that something can fall forever in a finite physical domain. Indeed, the bullet does not actually stop within a finite amount of time; it falls forever, and yet it travels only a finite distance of descent.

As the proposer points out in question 154, the answer lies in finding the sum of an infinite geometrical progression. In general, such progressions are a sequence of terms in which a constant (the first term in the series) is multiplied by a constant common ratio. Accordingly, the terms in the sequence will advance as: a, ar, ar^2, ar^3, ar^4, ..., ar^n, ... and so on, where a is a constant and r is the common ratio. If the sum of this sequence to the nth place is written as S_n, then it can be shown that: $S_n = \sum_{k=1}^{n} ar^{k-1} = a(1-r^n)/(1-r)$. Furthermore, provided r is smaller than 1, then as n becomes infinitely large, r^n approaches zero. In this manner, even after summing over an infinite number of terms a finite number $S_\infty = a/(1-r)$ is obtained. Returning now to question 154, we have $a = 20$, and $r = 19/20$, this is just the common ratio derived by dividing the first two terms. As a check on these quantities, the third term in the series will be given by the expression $20(19/20)^2 = 361/20 = 18 + 1/20$, as required — the fourth term will be: $17 + 59/400$. For the deduced a and r values to question 154, the sum to infinity is $S_\infty = 400$ miles. So, here is the answer to the Rev. Mr. T-H-'s question, over all eternity the bullet will fall just 400 miles — the bottomless pit isn't that deep after all, being equivalent to just 6.3% of the Earth's radius. The correspondence section of *Martin's Miscellaneous Correspondence* for

December 1757 informs us that the first correct answer was received from Mr. R. Peckham of Seal in Kent — a total of 12 correct answers to question 154 were received all together.

Although not required to answer the problem as posed, it is straightforward to determine the velocity and deceleration attributable to the infinitely falling bullet. Indeed, the velocity at the nth step will be $V_n = 20(19/20)^{n-1}$ miles per minute, while the acceleration or change in velocity between the nth and $(n+1)^{th}$ step is $D_n = -(19/20)^{n-1}$ miles per minute squared, where the minus sign indicates that the velocity is slowing down. Accordingly, as n becomes larger and larger, both the velocity of decent and the deceleration become smaller and smaller, tending after an infinite amount of time towards zero. This situation of describing smaller and smaller intervals for ever and ever reminds us of the classic conundrum of Zeno's paradox.

Zeno's paradox is a mathematical problem [22], rather than an actual, real-world physical problem, so too, in similar vein, is question 154 by "Rev. Mr. T-H" entirely predicated upon the mathematical principles of an unattainable infinity. In this manner the question, as posed, makes no physical sense and no such motion for a falling bullet could be engineered in the real world. This being said, it is entirely possible that an object might be made to fall a finite distance in a finite amount of time, before coming to a hovering stop for the rest of eternity (or for at least a very long time). Such a situation is, in fact, exactly that encountered by a golf ball falling through an Earth tunnel that has not been isolated from the Earth's atmosphere — as discussed in Chapter 17.

C.P. No. 1's solution

The all-important constant-density condition that makes Cymro's problem analytically tractable is outlined in Chapters 3, 17 and 18. The equation of motion for the tunnel-crossing particle of mass m is obtained from Newton's third law, which gives the acceleration $a = d^2r/dt^2 = F_{gravity}/m = -GM(r)/r^2$, where $M(r)$ is the mass interior to radius r. Given a constant density ρ, the mass interior to radius r can be written as $M = (4\pi r^3/3)\rho$, and consequently $d^2r/dt^2 = -Kr$, where $K = (4\pi/3)G\rho$ is a constant. The solution to this second-order differential equation can be found by substitution and by setting $r = R \cos(\omega t)$, where R is the radius of the Earth and ω is a constant, we have:

$$r = R\cos(\omega t)$$

$$\frac{dr}{dt} = -R\omega \sin(\omega t)$$

$$\frac{d^2 r}{dt^2} = -R\omega^2 \cos(\omega t) = -\omega^2 r$$

where we have the boundary conditions that at time $t = 0$, $r = R$ and the initial velocity $V = dr/dt = 0$. From the above equations we deduce that $\omega^2 = K = (4\pi/3)G\rho$, and this quantity, through the sine and cosine terms, gives the period of oscillation as $T = 2\pi/\omega$. A further simplification can be made with respect to ω by noting that Earth's surface gravity is given by the expression $g = GM(r)/R^2$, and that the density $\rho = M(r)/(4\pi R^3/3)$, which combine to give, $\omega^2 = g/R$. The period of oscillation is therefore $T = 2\pi(R/g)^{1/2}$, and this indicates that the Earth-crossing time is $t_{tunnel} = \pi(R/g)^{1/2}$. The maximum speed of the tunnel-crossing particle is realized when $\omega t = \pi/2$, at which instant, $V_{max} = dr/dt = R\omega = (gR)^{1/2}$ and at this moment the acceleration is zero. The time variation of the r, dr/dt and d^2r/dt^2 terms are shown in Figure 3.2.

Across a hollow Earth

Here is a problem set by Peter Tait and William Steele in the 1856 text, *A Treatise on the Dynamics of a Particle*. The problem reads: "The Earth being supposed a thin spherical shell, in the surface of which a circular aperture is made, if a particle be dropped from the center of the aperture, determine its velocity at any point of the descent." To begin, let the Earth's spherical shell have radius R and the particle a radius r. When the particle is held in place above the aperture it will experience a gravitational force due to the Earth's gravitational attraction, with $F_{out} = GMm/R^2 = gm$, where M and m are the mass of the Earth and particle respectively. In contrast once the particle has passed through the aperture, then Newton's shell theorem tells us that the accelerating force will drop to zero: $F_{in} = 0$, hence the velocity with which the particle will move on the inside of the shell is simply that which it gains in falling through its own diameter through the aperture. During this fall time, the shell being thin, the acceleration can reasonably be taken as being constant, and accordingly, the velocity of the particle upon entering the interior of the shell is given by Newton's laws of motion as $v^2 = 2a(2r)$, where a is the acceleration. Newton's second law indicates that $a = F_{out}/m = g$. Accordingly, the particle velocity at any point inside of the shell will be $v = 2(rg)^{\frac{1}{2}}$ — in this case, the velocity will be dependent upon the size r of the particle, since the motion is determined by the time it takes the particle to fall through the aperture. The travel time of the particle across the shell-Earth's diameter will be $t_{shell} = 2R/v = R/(rg)^{\frac{1}{2}}$. If we let the particle fall from *infinity* before it enters the shell, then the velocity of the particle will correspond to the shell's surface escape velocity: $v_{esc} = (2gR)^{\frac{1}{2}}$. The shell-crossing time accordingly becomes $t_{shell} = 2R/v_{esc} = (2R/g)^{\frac{1}{2}}$. Under the in-fall-from-*infinity* condition the shell-crossing time is $t_{shell} = (2^{\frac{1}{2}}/\pi) t_{tunnel} = 19.0$ minutes. The Earth-crossing time is remarkably short in the long in-fall case since the velocity of the particle $v_{esc} = 11.2$ km/s is both constant throughout the shell's interior and higher than that achieved in the Earth-tunnel model, where V_{max} at Earth's center only reaches 7.9 km/s.

The baron's hollow Moon and Earth

Hugo Gernsback in his *The Scientific Adventures of Baron Munchausen* (1915) introduced the idea that the Moon is a hollow shell (recall Chapter 9). Motion through such an object is a hybrid exercise set between that of the thin-shell problem and the full Earth-tunnel problem. Let the radius of the shell world be R and the depth of the rim be Δ (Figure A.2).

The gravitational force acting upon a body at a distance r from the center is now,

$$g(r) = -G\frac{4}{3}\pi\rho\frac{\left(r^3 - \delta^3\right)}{r^2}, \qquad \text{for } \delta < r < R,$$

$$g(r) = 0, \qquad \text{for } r < \delta,$$

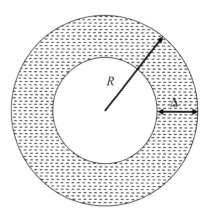

Figure A.2.   The hollow-Earth model. The region between $(R - \Delta)$ and the center is considered to be composed of empty space.

where $\delta = R - \Delta$. For a particle moving through the shell the equation of motion will be

$$\frac{d^2r}{dt^2} = -K\left(r - \frac{\delta^3}{r^2}\right),$$

where $K = \frac{4}{3}\pi G\rho$. When the particle is moving within the hollowed-out core (i.e., when $r < \delta$), the acceleration is zero: $d^2r/dt^2 = 0$. In the case that $\Delta \to R$, $\delta \to 0$, and the equation of motion is that for a uniform sphere (i.e., C.P. No. 1's solution). In the case where $\Delta \to 0$, we recover the thin-shell problem with $d^2r/dt^2 = 0$ at all times, indicating a constant fall velocity. While the two extreme conditions provide for straightforward analytic solutions, no such luxury exists for the in-between case. Accordingly the equation of motion will need to be solved for numerically with an appropriate set of boundary conditions: i.e., at $t = 0$, $r = R$, $v = 0$. A set of solutions to Hugo's Moon-shell problem have been obtained, and taking $R = 1{,}738$ km and $\Delta = 805$ km it is found that the travel time across the Moon and back again is 124 minutes — just 17 minutes longer than if the Moon were solid through and through (for which the period of oscillation is 106.83 minutes).

Numerical solutions for the fall time to the center of a hollow Earth, for varying shell thicknesses, have also been obtained, and the journeys

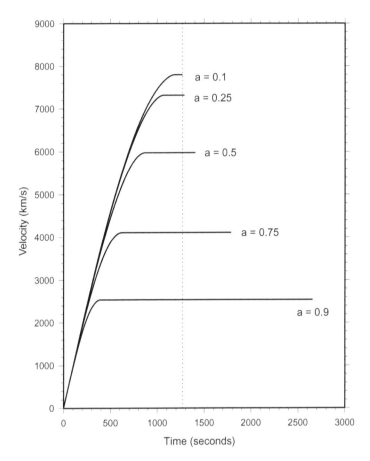

Figure A.3. Hollow-Earth velocity and travel times plotted according to shell thickness — the curves are labeled according to the normalized value of the shell's inner radius a, and the various curves end at the time of passing through the Earth's center. The dashed vertical line indicates the time to reach the center in the monolithic Earth situation (see Figure 3.2).

are illustrated in Figure A.3. For these simulations we have adjusted the density term entering into constant K so that the acceleration due to gravity at the surface is always $g = 9.81$ m/s^2, and the shell thickness is parameterized according to its inner edge a being some fraction of the Earth's radius, with $a = 0.9$ corresponding to a very narrow, very high density shell, and $a = 0.1$ corresponding to a near-monolithic globe. The travel time to the center with $a = 0.9$ turns out to be about twice that of the $a = 0.1$ travel time, since most of its journey to the center is made at a relatively low, constant speed of 2.5 km/s.

Terricola's prize questions

Both of Maskelyne's prize questions to the *Ladies' Diary* were answered by *Amicus* (meaning *Friend*), the pseudonym of the Reverend Charles Wildbore; a highly accomplished mathematician and longtime Editor (from 1780 to 1802) of the *Gentleman's Diary, or the Mathematical Repository*. Wildbore is largely recognized for his review work relating to mathematical papers submitted to the Royal Society of London, and for his many contributions and answers to mathematical questions posed in Benjamin Martin's *Miscellaneous Correspondence in Prose and Verse* (published between 1759 and 1764). This latter correspondence was bound into an impressive four-volume set of books, totaling over 2,000 pages of illustrations and text. A search through Martin's *Miscellaneous* revealed no hints of a question similar to Cymro's problem, but it did reveal the bottomless pit question (as described above) by the "Rev. Mr. T-H-", this latter question having resonance with the infinite fall time to the Earth's center as presented by the Merton calculators (recall Chapter 11).

The answers to the prize questions, as compiled by Leybourn, are to the modern eye somewhat obscure and a little difficult to follow, the solutions generally being presented with minimum explanation and very little discussion, and additionally, the methods and symbols employed are typically different to those that would be applied today. Below, however, I attempt to make the answers a little more accessible.

The 1795 prize question reads as: "Suppose the whole terraqueous globe taken as a sphere, should be instantaneously turned into a uniform elastic aeriform fluid, whose particles repel one another with a force which is to that which those of air repel one another, as the density of one to the density of the other, it will expand itself either to a finite or infinite extent, still preserving the form of a sphere. It is required to determine the force of gravitation tending towards the center, and also the density, at any given distance from the center; supposing the mean density of the Earth to be 3,825 times that of air at the surface of the Earth." Two correct answers were received for this question, one by *Amicus* (Rev. Wildbore) and another by *Clericus* of Southwold. *Clericus* was presumably a cleric, and ˎ Southwold is a small coastal town in Suffolk, England and the 1795 prize question is the only time that he contributed to the *Diary*. Unfortunately, *Clericus* was not an uncommon pseudonym of the times and the true identity of our erstwhile prize winner remains a mystery.

I have not been able to find any earlier versions of this prize question, and it does not appear to have had any extensive re-use. Indeed, the question is entirely abstract and involves a physical contradiction. However, let us proceed with *Amicus* and define the quantities to be used: *x* is the variable distance from the center of the aeriform sphere, *y* is the density at distance *x*, and *q* is the quantity of matter in the sphere of radius *x*. In modern terms *Amicus* begins by considering the conditions of hydrostatic equilibrium which is described according to the equation $dP/dr = -Gm\rho/r^2$, where *P* is the pressure at radius *r*, *m* is the mass interior to *r*, and *G* is the gravitational constant. In the normal sense of its application, hydrostatic equilibrium requires that the outwardly directed pressure force exactly balances the inwardly directed force of gravity; the two forces effectively finding a radius at which a dynamical stand-off can occur.

The question presented by Terricola allows for the substitution that $dP = d\rho/C$, where *C* is a constant to be determined — now, converting the symbols introduced by *Amicus* to modern form we derive the following initial result:

$$dy = -C\frac{q\,y}{x^2}dx \tag{A.1}$$

In addition, the amount of matter *dq* in a small shell of thickness *dx* will be

$$dq = 4\pi\,y\,x^2 dx \tag{A.2}$$

Using the condition that *dq* is constant, it can be shown that (A.1) and (A.2) combined yield (after a fair bit of algebra) the result that $q = 2x/C$ — which indicates that the mass interior to *x* increases as the distance *x* increases. Combining this result with (A.1) further yields the relationship: $y = 1/(2\pi Cx^2)$. From the information provided in the question we may now solve for the constant C, in that the density $y = 1/3825$ when $x = r$, where r is the radius of the Earth, and accordingly we have as our solutions:

$$y = \frac{r^2}{3825\,x^2} \quad \text{and} \quad q = \frac{4\pi\,r^2}{3825}\,x$$

These equations provide us with the answer that within the aeriform sphere the density varies as x^{-2}, the mass interior to x varies as x, and the force of gravity $F \sim q/x^2$ will vary as x^{-1}. The question allows for the infinite expansion of the aeriform sphere, but this would require that $q \to \infty$. If, however, we imagine that the dispersed particles of the aeriform sphere expand outwards to the limit which contains the Earth's mass, then the limiting radius will be $x_{lim} = 3825dr/3$, where d is the density of the Earth. In the solution provided by *Clericus*, the Earth's radius is taken to be 3,977 miles, and the density of the Earth is taken to be 3.5 times that of water: with these quantities given it is then argued that "$x = 1,440,000$ miles [is] the radius of the whole aerial sphere, containing matter equal to the whole Earth."

There is an interesting present-day analog to Terricola's 1795 prize question. If we ask what the circular velocity V of a particle would be at distance x from the center of the aeriform sphere then $V^2 = Gq/x =$ constant, since q varies linearly with x. In other words, the circular velocity at any location within the aeriform sphere is an invariant constant term. This situation has a parallel with the observed rotation velocity versus distance data pertaining to spiral galaxies (our own Milky Way galaxy included). Indeed, since the mid-1970s it has been clear that the rotation velocities of the stars and gas clouds within our galaxy do not decrease with increasing distance from the galactic center, but, with small variations, they remain essentially constant. The same constant velocity with increasing distance situation holds for all spiral galaxies that have ever been studied. This result is in complete contrast to expectation, since Kepler's 3rd law predicts that the rotation velocity should decrease as the inverse square root of the distance from the galactic center — this result follows from the fact that most of the observed stars reside in the galactic bulge. The constancy of galaxy rotation curves is generally explained in terms of dark matter — a gravitating component of the galaxy (and universe) that does not interact with electromagnetic radiation (and hence cannot be observed directly with any standard telescope). Indeed, dark matter appears to account for at least 25 percent of the mass within the universe — the ordinary matter component (the stars, interstellar and intergalactic gas) make up just 5 percent of the universe's mass budget. So, while dark matter is not directly observable (at the present time) and it is unclear what it actually might be (but it is usually taken as being some massive, currently unidentified elementary particle), its effect on galaxy rotation can be measured. Indeed, the simplest dark matter model that can account for the constant velocity rotation curves is that in which a galaxy sits at the center of a massive dark matter halo, with the amount of dark matter increasing linearly with distance from the center of the galaxy and the density of dark matter decreasing as the inverse distance squared.

While the Earth will never become an infinite aeriform sphere, the Sun and stars in general, to a certain extent can, and indeed, do. This corresponds to the so-called red giant phase of stellar evolution, when a star swells too many tens (even hundreds and thousands) of times its initial (so-called main sequence) size. The Sun during its red giant phase will swell to a size comparable to the Earth's orbit (that is expand by a factor in excess of 200), destroying and consuming Mercury and Venus in the process (and possibly the Earth). The reasons for why stars become red giants in their advanced age are still hotly debated, but one important contributing factor is that within their interiors some regions appear to behave as if they are $n = 5$ polytropes [44]. A polytropic gas is one in which the pressure P varies with the density ρ as $P = K\rho^{(1+1/n)}$, where K is a constant, and n is the polytropic index. For a star like the Sun, the interior structure is reasonably well explained by a polytrope of index $n = 3$. Inserting the polytropic relationship into the equations of hydrostatic equilibrium yields a set of relationships for the mass and radius of a star. Usefully, there are special, analytic solutions to the equations when the polytropic index n is set to 0, 1, or 5. The $n = 5$ solution is special in that the radius of the configuration tends to infinity, but the enclosed mass remains finite. Unlike the aeriform sphere, however, the mass interior to radius r and the density variation with r are somewhat complex. For very small radii, the amount of matter interior to r varies as r^3, but for very large radii the amount of matter interior to r becomes a constant.

Inversely, for very small radii the density at r is a constant, whereas for very large radii the density varies as r^{-5}.

The 1801 prize question by Terricola reads: "A quantity of matter being given, it is proposed to determine the figure of a solid of rotation made up of it, which shall have the greatest possible attraction on a point at its surface". Like the Earth-tunnel problem, this question has some considerable associated history. Indeed, in a slightly modified form, the same question can be found in Francis Holliday's *Miscellanea Curiosa Mathematica* (Volume 1) published in 1749. The question in this case was set under the pseudonym of Hurlothrumbo[1] and reads, "what form must a mass of homogeneous matter take, that the attraction thereof on a corpuscle, somewhere in or without its surface, may be the greatest possible. And what is the ratio of the attraction to the greatest whereby the corpuscle can be affected, when the same matter is of a sphere?" The answer to this problem was given under the record-breaking pseudonym of Philofluentimecanagegeomastrolongo. Indeed, our mysterious correspondent provides a masterful answer in the general case where the force of interaction varies as "the nth power of the distance" — the Newtonian gravitation case is recovered when $n = -2$. The same question was discussed in considerable detail by mathematician John Playfair in an 1812 article, *On the Solids of Greatest Attraction*, published in *The Edinburgh Philosophical Transactions*. In this work Playfair considers the question relating to the solids of greatest attraction, and notes in the very first paragraph of his article that "the investigations [...] were suggested by the experiments which have been made of late years concerning the gravitational of terrestrial bodies, first by Dr. Maskelyne on the attraction of mountains, and afterwards by Mr. Cavendish[2] on the attraction of leaden balls." Indeed, Playfair and geologist Lord Webb Seymor (10th Duke of Somerset) conducted a detailed geological survey of Schiehallion in 1811 in an attempt to better determine the specific density of the rocks out of which it is made. Playfair's solution to the solid of

---

[1] Hurlothrumbo was the title of a nonsense play written in 1729 by Samuel Johnson.

[2] This is the famous experiment conducted by Henry Cavendish in 1797–1798 in which the force of gravity between two masses in a laboratory was measured, leading to the first accurate value for the gravitational constant G.

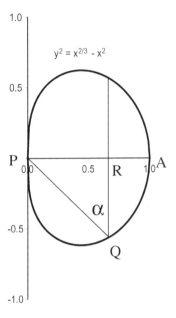

$$y^2 = x^{2/3} - x^2$$

Figure A.4. The profile that maximizes the gravitational attraction at point P — in answer to Terricola's 1801 prize question.

greatest attraction problem proceeds along the same lines as given in the *Ladies' Diary* (see below), but Playfair then extends the analysis to include the situation where the force of interaction varies as the distance to the nth power. Playfair further considers the situation when the particle is no longer located on the surface of the solid of greatest attraction, and he also determines the gravitational attraction associated with cylinders and with cones.

Solutions to Terricola's 1801 prize question were provided by *Amicus* and a Mr. John Ryley of Leeds. Ryley provides a diagram to illustrate his method of attack, and it is worth seeing the diagram for reference (Figure A.4). Imagine, Ryley argues, a particle Q located at some point on the curve which is to be found. Also, let the symmetry axes of the body be the line PA where the distance PA = a, which is a constant to be determined in terms of the mass of material m. Point P is the location at which the greatest possible gravitational attraction is to be realized. Ryley first notes that the gravitational force of a particle placed at Q will vary as the distance PQ^{-2} — this is just a statement of the inverse square law nature of Newton's formula for gravitational attraction. Next, Ryley notes that the gravitational force of the particle at Q in the direction PA will vary as $(PR/PQ)/PQ^2$, here $PR/PQ = \sin \alpha$, and accordingly, when angle $\alpha = 90$ degrees, Q, the location of the particle, will coincide with point A. When Q coincides with point P, then $\alpha = 0$ degrees, and there will be no gravitational force acting in the direction along PA. Now, by the wording of the question, Ryley notes that the gravitational attraction in the direction QR, which will vary as $(PR/PQ)/PQ^2$, cannot be larger than the gravitational attraction when the particle is located at point A. Accordingly, Ryley argues that when the whole gravitational attraction is at its greatest and at some constant value, so $(PR/PQ)/PQ^2 = PA^{-2}$. If we now introduce a bit of shorthand notation, and set $PR = x$, $QR = y$ and $PA = a$, then by Pythagoras's theorem, $PQ = (x^2 + y^2)^{1/2}$ and after a little algebra we have that $y^2 = a^{4/3} x^{2/3} - x^2$ and this is the profile of the curve PQA, and the bounding profile of the body of rotation (Figure A.4). In order to determine the value of $PA = a$ in terms of the given mass, we must find the volume V of the solid made by rotating the curve PQA about the axis PA. This last step entails evaluating the integral for the solid of rotation. Accordingly, we need to find

$$V = \pi \int_0^a f(x)^2 \, dx$$

and this can be related to the mass m and density ρ of the material by the definition that $\rho = m/V$. Since $f(x)^2 = y^2 = a^{4/3} x^{2/3} - x^2$ in our case, we have

$$V = \frac{3\pi}{5} a^{4/3} x^{5/3} - \frac{\pi}{3} x^3$$

which is to be evaluated between the limits $x = 0$ and $x = a$. Accordingly, $V = m\rho = (4\pi/15)a^3$, which gives a in terms of the given mass m and its density as $a = (15m\rho/4\pi)^{1/3}$ — in our diagram (Figure A.4) we have simply chosen the parameters m and ρ so that $a = 1$.

The tautochrone

The isochronal property of the tautochrone, in which the time of descent to the lowest point on a curve is equal and independent of the starting point, can be solved for in many different ways — the standard approach is to use Laplace transforms to find the solution. Here we shall simply assume that the tautochrone curve is the cycloid and work backwards to a proof that the motion must therefore be isochronal. The parametric equations for a cycloid are: $x = a(\varphi - \sin \varphi)$, and $y = a(1 - \cos \varphi)$, where a corresponds to the radius of the circle rolling along the straight edge, and $0 \le \varphi < 2\pi$. The motion to be considered is that of a bead moving along an inverted (that is U-shaped) cycloidal wire — with y now being parameterized as $y = a(1 + \cos \varphi)$, so that the starting point is at $(x_0, y_0) = (0, 2a)$, and the center and lowest point on the curve is at $(x, y) = (\pi, 0)$. The speed of the bead will be given by the time derivative of the distance s moved along the wire, with $v = ds/dt$. Following Norwegian mathematician Niels Henrik Abel, who produced the solution approach to be followed in 1823, we now invoke the conservation of energy on the bead, and this provides a relationship between its kinetic and gravitational potential energy — accordingly: $\frac{1}{2} mv^2 = mg(y_0 - y)$, where m is the mass of the bead, g is the acceleration due to gravity, $y_0 = 2a$ is the starting height and y is the vertical distance through which the bead has fallen. Simplifying the expression for the energy, we have:

$$v^2 = \left(\frac{ds}{dt}\right)^2 = 2g(y_0 - y)$$

and what we need to do next is find an expression linking the vertical height variable y to the path length variable s. This can be done through the differential expression for the path length: $(ds)^2 = (dx)^2 + (dy)^2$. Since we are assuming that the inverted cycloid satisfies the tautochrone solution, we can determine the differentials for dx and dy to reveal,

$$ds = \sqrt{1 + \left(\frac{dx}{dy}\right)^2}\, dy = \sqrt{1 + \left(\frac{1 - \cos\varphi}{\sin^2\varphi}\right)^2}\, \frac{dy}{d\varphi}\, d\varphi = -a\sqrt{2(1 - \cos\varphi)}\, d\varphi$$

This expression can now be integrated so that

$$s(y) = -a\sqrt{2} \int_0^\varphi \sqrt{(1 - \cos\varphi)}\, d\varphi$$

We now employ a few tricks — using the identity sin $(\varphi/2)$ = $[(1 - \cos\varphi)/2]^{1/2}$ the integral for $s(y)$ reduces to

$$s(y) = -2a \int_0^\varphi \sin(\varphi/2)\, d\varphi = 4a\cos(\varphi/2) + C$$

where C is a constant which for the present we take to be zero so that $s_0 = 0$ — by setting C to some non-zero value the arbitrary starting height condition for the tautochrone can be expressed. Using yet another trigonometric identity we can perform another substitution to reveal: $\cos(\varphi/2) = [(1 + \cos\varphi)/2]^{1/2} = [y/2a]^{1/2}$, and we accordingly recover the result that $y = s^2/8a$. The 8a term is, in fact, Christopher Wren's rectification result for the arc length of a cycloid generated by a wheel of radius a [see Chapter 18]. We can now substitute this identity back into the equation for the conservation of energy, and recover the result that:

$$\left(\frac{ds}{dt}\right)^2 = 2g(y_0 - y) = \frac{g}{4a}(s_0^2 - s^2)$$

and this expression we can differentiate with respect to time, to yield

$$2\left(\frac{d^2s}{dt^2}\right)\left(\frac{ds}{dt}\right) = -\frac{g}{2a}s\left(\frac{ds}{dt}\right)$$

and, finally, a little more algebra eventually reveals the equation of simple harmonic motion:

$$\left(\frac{d^2 s}{dt^2}\right) + \frac{g}{4a} s = 0$$

The latter equation shows that the bead will oscillate back and forth along the cycloid with its path length s varying as $s = s_0 \cos[(g/4a)t]$, and the time for the bead to complete one passage along the wire will be $T = 2\pi(a/g)^{1/2}$. In this derivation it is important to note that the equation for simple harmonic motion is exactly satisfied. This is in contradistinction to the fixed-length pendulum where the equation for simple harmonic motion is only satisfied in the small-angle-of-oscillation approximation. Showing that the cycloid also satisfies the brachistochrone, fastest descent time, condition between two points (under constant gravity) proceeds in a similar vein to that for the tautochrone. The mathematical details, however, require the use of the calculus of variation, which we shall not enter into here.

The self-tunneling black hole oscillator

The Cymro's problem in which an accreting black hole is imagined to be moving back and forth across the Earth's interior, accreting matter as it goes, has the equation of motion:

$$\frac{d^2 r}{dt^2} + 2b\frac{dr}{dt} + \omega^2 r = 0$$

which is the basic equation of motion that we have seen before (with $b = 0$), but now we have a damping term that depends upon the velocity dr/dt. The general solution to such a differential equation is

$$r = C_1 \exp(n_1 t) + C_2 \exp(n_2 t)$$

where C_1 and C_2 are constants determined by the boundary conditions and n_1 and n_2 are the roots corresponding to the auxiliary equation

$$n^2 + 2bn + \omega^2 = 0$$

such that

$$n = -b \pm \sqrt{b^2 - \omega^2}$$

In the case where $b^2 < \omega^2$, both n_1 and n_2 are imaginary and the solution to the differential equation becomes

$$r = R \exp(-bt) \cos(\alpha t)$$

where we have applied the boundary condition that at $t = 0$, $dr/dt = 0$ and $r = R$. The constant α is given by the relationship: $\alpha^2 = \omega^2 - b^2$. For the solution of Equation (20.1) we have the substitution that $b = \frac{1}{2}\beta\omega$, which gives $(dm/dt)/m = \beta\omega$ as the accretion rate. The e-folding time, during which the oscillation radius decreases by a factor of e, is thus $T_{ef} = 1/b = 2/\beta\omega$, and the oscillation period is $T = 2\pi/\alpha$.

There are further solutions to the equation of motion. If $b > 1$ or $b > \omega$ then n_1 and n_2 are both negative and the solution corresponds to a strongly damped motion described by two time-dependent decreasing exponential terms. In this case the motion may take the particle through the $x = 0$ central location once. If $b = 1$ or $b = \omega$, then $n_1 = n_2$ and a critical damping situation arises, with the motion of the particle being driven directly to $x = 0$ with no oscillations about the center taking place.

Notes and Selected References

[1]   From the very outset it probably makes for a useful exercise to try to distinguish between myths, rumors, a good story, and conspiracy theories. Galileo's supposed Leaning Tower of Pisa experiment is an example of a *good story* — it has the ring of truth to it, and whether it actually happened or not is not actually important. A *rumor*, in contrast, is a story, or tall tale, invented specifically to misinform (or, more harmlessly, to misguide) and it is typically based upon unverifiable information. In the modern era, another name for rumor would be fake news. *Myths* are somewhat similar to rumors, although unlike the latter, the former have some element of a lost truth buried deep within them. It is an historical myth for example that Pythagoras was the first person to prove the Pythagorean Theorem — this is known not to be the case, but the myth that he was the first to prove the theorem has perpetuated throughout history. *Conspiracy theories* are generally predicated on ignorance in the true meaning of the word: a lack of knowledge and/or information. Indeed, it is a sad indictment of modern society that conspiracy theories abound. There are, for so it seems, a vast multitude of people who hold the paranoid belief that their government or government agencies deliberately conspire to quash and conceal knowledge of miraculous events and/or otherworldly happenings (UFO landings and so forth). Not only this, so the conspiracy theorists will tell us, governments also set out to deliberately fool us, and for example such groups have claimed that the Apollo Moon landings never, in fact, took place, but were produced on some secret film stage, at some secret location, with a secret crew of actors and technicians (who never blabbed a word of what they did). The claims of a conspiracy are, of course, totally ludicrous and are completely based upon an ill-informed, ill-educated, poorly-reasoned world view. With respect to the Moon landings, one might, at best, argue that the

conspiracy argument is at least harmless; it has no effect upon anyone's life. For other topics, however, as in the case of the, so-called, climate change conspiracy, and/or the various vaccinations (as a means of governmental control) conspiracy, the arguments are far from harmless, and indeed, they are the cause of actual death and destruction. For a highly recommended discussion concerning some of the inherently illogical aspects of conspiracy theories, see the article *On the Viability of Conspiratorial Beliefs*, by David Grimes (Oxford University), published in the journal *PloS ONE* (January 26, 2016).

[2] The first issue of *The English Mechanic. A record of mechanical inventions, scientific and industrial progress, building engineering, manufacturers and arts*, was published on March 31st, 1865. Appearing weekly, the magazine cost just 1 penny and was edited by Passmore Edwards (who had previously been involved with the *English Mechanics* magazine, first published on October 19th, 1823). In early 1866 the title of the magazine was shortened to *The English Mechanics and Mirror of Science and Art*, and practically doubling in size, the price per issue was increased to 2 pence. Importantly in 1866 the magazine also opened a new section called the Subscriber's Exchange Club — a column that allowed readers to exchange articles without paying page charges. The *Mechanics* additionally expanded at this time its Letters to the Editor column. These columns eventually evolved into the Queries and Replies to Queries sections that made the *Mechanics* so popular with its readers. Writing just a few years before the *Mechanics* folded production, D. J. Smith (*English Mechanics* for January 18th, 1924) noted that "many of the leading scientists and engineers of today acknowledge their debt to the *English Mechanics* from an educational point of view." Among its more famous regular contributors were Lord Grimthorpe (Edmund Beckett — lawyer, renowned horologist and architect), Lord Rayleigh (John Strutt — 1904 Nobel Prize for Physics), Oliver Lodge (Professor of Physics and Mathematics, University College in Liverpool) and Oliver Heaviside (electrical engineer and mathematician) who used the simple *nom de plume* of O. Becoming increasingly paranoid and eccentric in later life, however, Heaviside was also known for adding the letters W.O.R.M. to his signature. He would also on occasion change his address from "Home Field," Torquy (in Devon, England) to "Worm Field," Torquay. In other letters he would add strange cryptic notes and/or messages. In one letter dated October 8th, 1922, Heaviside numbered the letters in his signature, which if one writes them out in the numbered sequence reads, "O He is a very devil W.O.R.M" (to make this

rearrangement actually work the first *i* in Heaviside was overwritten with a *y*). In actuality Heaviside could have added the much more prestigious F. R. S. (Fellow of the Royal Society) to his signature. Strange signatures and *nom de plumes* of one form or another were commonly used in the late 19th and early 20th century, and, for example, Nikola Tesla is known for adding the post-nominal letters G.I., for "great inventor," to his signature.

[3] The idea that what we call reality is in fact a complex interrelationship between at least three world domains (a Platonic-like ideal domain, the real world around us, and the cognitive world that we as human beings experience) has been discussed by Roger Penrose in his highly insightful book, *The Emperor's New Mind: Concerning Computers, Minds, and the Laws of Physics* (Vintage Books, 1989). I would also highly recommend to the reader a second Penrose book, *The Road to Reality: A Complete Guide to the Physical Universe* (Vintage Books, 2005).

[4] Even though one proof is enough to establish a mathematical theory, the theory of Pythagoras has been proven, by different methods, many, many times. A collection of some of these proofs is given in the book, *The Pythagorean Theorem: A 4,000-Year History* by Eli Maor (Princeton University Press, 2007). Mark Levi in his book *The Mathematical Mechanic: Using Physical Reasoning to Solve Problems* (Princeton University Press, 2009) provides a pleasing physical proof of the Pythagorean Theorem in terms of the conservation of energy. Imagine, he argues, that you are standing on a sheet of ice with perfectly frictionless shoes on (standing might be a little difficult — but this is a thought-experiment proof). You push off from one wall (the *x*-axis) at speed *a* along another wall (the *y*-axis, perpendicular to the *x*-axis). Your kinetic energy will therefore be $\frac{1}{2} ma^2$. After a while you push off from the *y*-axis with speed *b* along the direction of the *x*-axis, gaining extra kinetic energy $\frac{1}{2} mb^2$. Your kinetic energy after the second burst is $\frac{1}{2} m(a^2 + b^2)$. Your final speed *c*, however, after the second push will be along the hypotenuse of the velocity triangle with kinetic energy $\frac{1}{2} mc^2$. Conservation of energy now requires that $\frac{1}{2} m(a^2 + b^2) = \frac{1}{2} mc^2$, and after cancellation of the $\frac{1}{2} m$ terms the Pythagorean relationship appears.

[5] From Newton's third law, during a collision between two particles, the force F_1 exerted on particle 1, will be equal in magnitude and opposite in direction to the force F_2 experienced by particle 2 — that is $F_1 = -F_2$. Defining the momentum *P* of a particle to be the product of its mass *m* multiplied by its velocity *v* — such that $P = mv$ — Newton's second law now relates the force *F* to the acceleration *a* through the change in velocity

dv/dt and the rate of change in the momentum, with $F = ma = m(dv/dt) = dP/dt$. From this latter relationship linking the force to the rate of change of momentum, Newton's third law can be rewritten as: $F_1 = dP_1/dt = -F_2 = -dP_2/dt$, and accordingly $d(P_1+P_2)/dt = 0$, and this tells us that the sum of momenta $P_1 + P_2$ is a conserved quantity, that is, the total momentum does not change during an interaction.

[6]   The escape velocity V_{esc} can be derived upon a straightforward application of the conservation of energy, and it corresponds to the minimum velocity with which a projectile must be fired from the Earth's surface in order to just escape its gravitational influence at "infinity" — that is, at a great distance from the Earth's surface. At the Earth's surface the kinetic and gravitational potential energy will be $\frac{1}{2} mV_{esc}^2 - GMm/R$, where *m* is the mass of the projectile, and where *M* and *R* are the mass and radius of the Earth. The velocity of the projectile at *infinity* (or more precisely at some arbitrarily great distance *D* away from the Earth, with $D \gg R$), will be essentially zero, and likewise, so too will the gravitational potential energy be zero (since, $0 \approx GMm/D \ll GMm/R$ for $D \gg R$). The total energy at *infinity* will therefore be of order zero. The conservation of energy therefore dictates that $\frac{1}{2} mV_{esc}^2 - GMm/R = 0$, and accordingly: $V_{esc} = (2GM/R)^{\frac{1}{2}}$. For the Earth, the escape velocity is 11.186 km/s; for the Moon it is 2.38 km/s; for the Sun it is 617.5 km/s.

[7]   The key reaction being measured was that of the neutrino interaction with chlorine, whereby: $v_e + {}^{37}Cl \rightarrow {}^{37}Ar + e^-$. After a specified exposure time the target fluid is *filtered* in order to determine the amount of ${}^{37}Ar$ that has been produced, and this, with an estimate of the number of target ${}^{37}Cl$ atoms in the detector fluid, provides a measure of the neutrino flux from the Sun.

[8]   "Shew that the velocity acquired by a body in falling from infinity to the Earth's core, is to the velocity of a secondary at the Earth's surface as $\sqrt{3}:1$." The answer to this question is built around the concept of the conservation of energy, that is the sum of the potential energy *U* and the kinetic energy *K* is constant: $U + K$ = constant. Assuming that the particle has zero initial velocity and is at a large (that is an *imagined* infinite) distance from the Earth then the initial energy is zero, which indicates that $U + K = 0$ at all times throughout the particle's flight. For the particle stopping at the Earth's surface, we have $\frac{1}{2} mv^2 - GMm/R = 0$, where *M* is the mass of the Earth and *R* is its radius, *G* is the universal gravitational constant, *m* is the particle mass and *v* is the impact velocity. Accordingly, $v^2 = 2GM/R$. This is the escape velocity derived earlier in [6]. For a particle falling all the way

to the Earth's center, the gravitational potential is taken to be that of a uniform sphere of constant density and radius R, with $U(r) = -(GM/2R^3)$ $(3R^2 - r^2)$, where r is the distance from the center. At the center, therefore, the velocity u will be $u^2 = 3GMm/R$. For a particle starting at Earth's surface and falling to the center, however, the initial velocity is zero, while the initial potential energy is $-GMm/R$. Accordingly, the velocity that this particle will have at the center is $v^2 = GM/R$. We now have the required result that $u^2/v^2 = 3$.

[9] For an excellent biography account, see Derek Howse, *Nevil Maskelyne: The Seaman's Astronomer* (Cambridge University Press, 1989).

[10] See Edwin Danson, *Weighing the World: The Quest to Measure the Earth* (Oxford University Press, 2006).

[11] See Derek Howse, *Greenwich Observatory.* Volume 3: *The Buildings and Instruments* (Taylor and Francis, 1975).

[12] For a comprehensive review see David W. Hughes, "On seeing stars (especially up chimneys)," *Quarterly Journal of the Royal Astronomical Society,* **24**, 246–257 (1983). See also, Alex Smith, "Daylight visibility of stars from a long shaft," *Journal of the Optical Society of America,* **45**(6), 482–483 (1955). An excellent account of atmospheric seeing, rainbows, halos, and coronae, is given by M. G. J. Minnaert, *Light and Color in the Outdoors* (Springer-Verlag, 1974).

[13] Stellar parallax is a direct means by which the distance to a star can be measured according to the scale of the Earth's orbit about the Sun = 1 AU = 1.495×10^{11} meters. The method is based upon measuring the apparent shift in the position of a nearby star, relative to much more distant stars, over a six-month time interval, as the Earth shifts in position through the diameter of its orbit about the Sun. If angle P is half the measured angle of shift over a six-month time interval, then the distance to the star will be d (AU) $= 1/\tan P$. Since the angle of stellar parallax is invariably very small, a small-angle approximation to the tangent operator can be made, and accordingly stellar distances are determined according to the formula d (pc) $= 1/P$ (arc seconds), where the parsec (pc) is defined as being the distance when the angle of parallax $P = 1$ arc second.

[14] A very detailed and informative biography of Hooke has been given by Stephen Inwood: *The Man who Knew Too Much* (Macmillan Pub. Ltd., 2002). Also highly readable is *The Diary of Robert Hooke, M.A., M.D., F.R.S., 1672–1680*, edited by H. W. Robinson and W. Adams (Taylor & Francis, 1935).

[15] G. B. Airy, *Lecture on the Pendulum-Experiments at Harton Pit* (Longman & Co. 1855). See also [10] and [18].

[16] The number of swings that a pendulum of length l will make per day is $N = 1\ \text{day}/T$, where $T = 2\pi(l/g)^{1/2}$. For two identical-length pendulums, one located at the top of the mine and the other at the bottom, the ratio of the number of swings per day will be $N_{top} = N_{bottom}(g_{top}/g_{bottom})^{1/2}$, where the subscripts indicate the pendulum locations. In a homogeneous sphere, $g_{top} > g_{bottom}$ and accordingly we would have $N_{top} > N_{bottom}$.

[17] For a detailed history of the measurement of Earth's mean density, see the article by David W. Hughes, "The mean density of the Earth," *Journal of the British Astronomical Association*, **116**, 21–24 (2006).

[18] See the author's text, *The Pendulum Paradigm: Variations on a Theme and the Measure of Heaven and Earth* (Brown Walker Press, 2014).

[19] Johannes Kepler deduced many laws and harmonies with respect to planetary motion and orbital spacing. With respect to Tycho Brahe's data on the sky positions of Mars, however, he derived what are now known as his three laws of planetary motion. The first law states a planet's orbit is elliptical in shape with the Sun being located at one of its focal points. The second law indicates that equal areas are swept out by the Sun–planet line in equal intervals of time. The third law provides a numerical relationship between orbital period P and the semi-major axis a of a planet's orbit: with $P^2 = Ka^3$, where K is a constant.

[20] The inspiration behind my flight recording experiment was from a reading of *The Science of Disk World IV: Judgment Day*, by Terry Pratchett, Ian Stewart and Jack Cohen. This wonderful tome (part of an excellent series of books) in a round-about-way also introduces the reader to the conundrum, "what kind of bridge is made of nothing but thin air?" To which the answer is "a tunnel."

[21] The author verified the great power of Eratosthenes's method for determining the size of the Earth on June 20th, 2004. In my case the sundial experiment was conducted from two locations in the prairie province of Saskatchewan, Canada, with one sundial set-up in Regina and a second sundial placed in Star City (287.8 ± 0.2 km due north of Regina). The sundial shadow lengths from the two locations were found to be 2.62 ± 0.12 degrees different (for simultaneous observations), giving a measure of the Earth's radius of 6,294 ± 290 km. For details, see "The measure of the Earth: a Saskatchewan diary," *Journal of the Royal Astronomical Society of Canada*, **99**, 7–9 (2005).

[22] There are a number of different forms of Zeno's paradox, but they essentially all revolve around the idea of the continued division of an interval of space and/or time. Zeno asserted that between any two points in space (or

time) there is always another point, and accordingly to travel from some point A to another point B one must move through an infinite number of intervals. The most common version of the paradox is that expressed in terms of a supposed race between the extremely athletic Achilles and a slow-moving tortoise that has been given a head start towards the finish line. The story then reveals that whenever Achilles arrives at a point where the tortoise has already been, the tortoise will have always moved a little further onward. Accordingly, even though the distance that Achilles has to cover in order to catch the tortoise is getting smaller and smaller, there are nonetheless an infinite number of intervals that need to be crossed. This reasoning expanded into the concept that all motion must be illusory. In the real world, however, space and time are not infinitely divisible and there are practical limits to the division of any interval — this effectively removes Zeno's assumption that there is always a mid-way point to any interval. That 4-dimensional spacetime is fundamentally discreet, that is not infinitely divisible, is effectively realized through the units of the Planck length ($\sim 10^{-35}$ m) and Planck time ($\sim 5 \times 10^{-44}$ s) — these being *natural* units of length and time as first established by physicist Max Plank in 1899. Infinities such as that described by Zeno are essentially mathematical problems, rather than issues of physical impossibility. In this latter sense, an infinity that can exist in the Platonic (mathematical) idealized world does not cross over into physical actuality and the experienced (that is real) world. See also [3].

[23] Galileo provides a beautiful description of the idea of relativity in his 1632, *Dialogue Concerning the Two Chief World Systems*. Galileo argues, "Shut yourself up with some friends in the main cabin below decks on some large ship, and have with you there some flies, butterflies, and other small flying animals. Have a large bowl of water and some fish in it. Hang up a bottle that empties drop by drop into a wide vessel beneath it. With the ship standing still, observe carefully how the little animals fly with equal speeds to all sides of the cabin. The fish swim indifferently in all directions; the drops fall into the vessel beneath... When you have observed all these things carefully, have the ship proceed with any speed you like, so long as the motion is uniform and not fluctuating this way and that. You will discover not the least change in all the effects named, nor could you tell from any of them whether the ship was moving or standing still." This story illustrates the idea of Galilean relativity, a concept that lies at the very core of the Copernican model for the description of the solar system. It is this principle that allows for the Earth to be in motion about

the Sun without us, as observers on the Earth, having any sense of its motion. Indeed, the Earth (along with you and me) is traveling at some 30 km/s around the Sun.

[24] This statement involves a little algebra in its proof, and is centered on eliminating the time t parameter from the equations for x and y. It can be shown that $y = ax + bx^2$, where a and b are constants, and this is the equation for a parabola. The equation describes a parabola with its vertex tilted upwards and symmetrical about the vertical line through $x = \frac{1}{2} R_{max}$. The constants are given by: $a = \tan \alpha$, and $b = -\sec^2\alpha/(2gR_{max})$.

[25] While the fundamental constants, such as the speed of light, Planck's constant and the gravitational constant, are generally taken to be constant over time, there is still the open possibility that they do in fact change on time scales comparable to the age of the universe. One modern-day example of a fundamental constant that may change with time (that is the age of the universe) is that of the cosmological constant Λ which describes the energy density of the vacuum of space. Various, largely disputed experiments also suggest that over the age of the universe there have been variations in the speed of light, the gravitational constant and the value of the fine structure constant. For further reading, see John Barrow, "Cosmology: A matter of all and nothing" (*Astronomy and Geophysics*, **43**, 4.8-4.15, 2002), and John Webb, "Are the laws of nature changing with time?" (*Physics World*, April 2003). For a detailed technical review see Takeshi Chiba, "The constancy of the constants of Nature: Updates (http://arxiv. org/abs/1111.0092).

[26] Voltaire published a second work, called *Micromégas*, in 1752, in which he continued to lampoon Maurpertuis and his highly successful expedition to Lapland. Micromégas is a giant from a planet orbiting the star Sirius, who along with a dwarf that he befriends from planet Saturn, visits Earth just in time to catch the French expedition as it is sailing home through the Baltic Sea. The novella is generally taken to be a satirical account of human hubris and idiosyncrasy. Indeed, taking pity upon the tiny humans ("intelligent atoms") Micromégas presents the expedition's philosophers with a book that "contain[s] the point of everything"; the book is dutifully taken to Paris, but when the book is opened by the "ancient secretary" of the Académie all that is found are blank pages — "Ah!" he said, "I suspected as much." The abrupt ending of *Micromégas* leaves much to the imagination, and it is often referred to as *Voltaire's Riddle*. In similar vein to Micromégas's empty book, British astrophysicist Arthur Eddington worked according to the philosophical challenge of the "blank page." Indeed, he reasoned that

the fundamental workings of the universe could be understood and expressed through pure reasoning and mathematics alone. Within this outlook every new blank page encountered within a notebook can be thought of as a portal for the exploration and discovery of the universe around us. Eddington developed his ideas in a series of lectures delivered at Trinity College, Cambridge in 1938, and in his subsequent book, *The Philosophy of Physical Science* (Cambridge University Press, 1939).

[27] Here I have used Ian Bruce's translation of Euler's *Mechanica*, which can be found at http://www.17centurymaths.com/.

[28] Association Française pour L'Avancement Des Sciences, Compte Rendu De La 11 Session, La Rochelle, 1882. See http://gallica.bnf.fr/ark:/12148/bpt6k201158q/f236.image. Collignon published a number of texts on railway engineering, mechanics and the theory of dynamics. Perhaps Collignon's best known work, however, is his multi-volume *Traité de Méchanique*, first published between 1873 and 1874. A review of these writings revealed no prior consideration of the Earth-tunneling problem.

[29] The balance point of a Bullet Train: The N700 Series Shinkansen (Bullet) train can transport passengers along surface hugging rails at speeds close to 300 km/hr. Its frontal cross-section area $A = 13$ m^2, and with 16 carriages attached the mass comes in at about $m = 73,000$ kg. The balance point, that is, the location within the Earth at which the gravitational force across the cross-section area of the train (imagined to be moving down a vertical tunnel) equals the upward pressure exerted by the air in the tunnel, is given by the condition that $g(r)A = P(z)$, where $g(r)$ is the gravitational force at distance r from the Earth's center and $P(z)$ is the air pressure at a distance z below the Earth's surface. Taking the Earth to be a sphere of constant density $\rho = 5,500$ kg/m^3, we can write $g(r) = (4\pi G/3)\rho m r$. Substituting for known numbers the balance point is established by the condition that $8.34 \times 10^7 (1 - z/R) = P(z)$, where R is the radius of the Earth. Using the formula derived for the barometric pressure $P(z)$ in Chapter 17, the balance condition is found to hold at $z \approx 54$ km below the Earth's surface. We have the result, therefore, that if we imagine gently lowering a 16-carriage bullet train down a vertical tunnel then at about 54 kilometers below Earth's surface the opposing pressure force of the air (at some 850 atmospheres) will fully support the downward-pointing train and it will just hover at this location moving no closer to the Earth's center. There are, as always, a good number of caveats to a result such as this, but probably the most important one is that at the pressures being envisioned the ideal gas equation will no longer hold true and so-called Van de Waals forces will need to

be taken into account. Even with this latter refinement, however, the pressure gradient will still rise to a level at which our downward-pointing train will be supported.

[30]   A standard 4th-order Runge–Kutta numerical integration scheme was developed to solve for Equation (17.4). The approach followed was to break down the second-order differential equation into a first-order one, with the transform: $d^2h/dt^2 = dV/dt$, where $V = dh/dt$. The numerical scheme then solves for the velocity as a function of time after the golf ball is released, with the distance d traveled in each time step evaluated as $d = V \times t_{step}$, where t_{step} is the time-step interval (chosen to be ½ second). In the situation where only contact friction is assumed the equation to solve for is $dV/dt = g(h) - f$, where f is expressed in newtons per kilogram — when $f = 0$, the simple harmonic solution appropriate to Cymro's problem is re-encountered.

[31]   See Paul Cooper, "Through the Earth in forty minutes," *American Journal of Physics*, **35**, 68–70 (1966). For a comprehensive mathematical account of the gravity-train problem (including the spin of the Earth), see A. J. Simoson, "Falling down a hole through the Earth," *Mathematics Magazine*, **77**, 171–189 (2004).

[32]   Kepler's problem is concerned with the determination of a planet's position within its orbit at some specified time t after a given perihelion passage at time t_{per}. Introducing the mean anomaly $M = n(t - t_{per})$, where $n = 2\pi/P$ and P is the orbital period. The eccentric anomaly E is then determined from the relation $M = E - e \sin E$, where e is the orbital eccentricity. With E determined the heliocentric distance r is given as $r = a[1 - e \cos E]$, where a is the orbital semi-major axis. The true anomaly ν, the positional angle of the planet as measured from the Sun to perihelion line, is further given by the relationship $\tan(\nu/2) = [(1 + e)/(1 - e)]^{\frac{1}{2}} \tan(E/2)$. Historically, Kepler's problem has revolved around the determination of the eccentric anomaly E from the mean anomaly M and the eccentricity e. There is no analytic solution to the requisite formula and it has to be determined iteratively. For a detailed historical account of Kepler's problem and its solution, see the author's book, *The Wayward Comet: A Descriptive History of Cometary Orbits, Kepler's problem and the Cometarium* (Universal Publishers, 2016).

[33]   There are many, many articles and books concerned with the solution of the brachistochrone and tautochrone problems. The ones that I have found particularly useful are: Ramesh Chander, "Gravitational fields whose brachistochrones and isochrones are identical curves," *American Journal of Physics*,

45, 848–850 (1977); John McKinley, "Brachistochrones, tautochrones, evolutes and tessellations," *American Journal of Physics*, **47**, 81–86 (1979); Amanda Maxham, "Brachistochrone inside the Earth: The gravity train," http://www.physics.unlv.edu/~maxham/gravitytrain.pdf; and R. Gómez, V. Marquina & S. Gómez-Aiza, "An alternative solution to the general tautochrone problem," *Revista Mexicana de Física E*, **54**(2), 212–215 (2008).

[34]   See A. Dziewonski & D. Anderson, "Preliminary Reference Earth Model," *Physics of the Earth and Planetary Interiors*, **25**, 297–356 (1981); a pdf copy of this paper is available at http://www.cfa.harvard.edu/~lzeng/papers/ PREM.pdf. See also A. Klotz, "The gravity tunnel in a non-uniform Earth," http://arxiv.org/abs/1308.1342.

[35]   D. Turcotte, and S. Emerman, "Dissipative melting as a mechanism for core formation," *Journal of Geophysical Research* (*Supplement*), **88**, 891–896 (1983).

[36]   See, e.g., M. I. Ozhovan *et al.*, "Probing of the interior layers of the Earth with self-sinking capsules," *Atomic Energy*, **99**, 556–562 (2005); and W. Chen, J. Hao & Z. Chen, "A study of self-burial of a radioactive waste container by deep rock melting," *Science and Technology of Nuclear Instillations*, **2013**, 1844757 (2013).

[37]   David Stevenson, "Mission to Earth's core: A modest proposal," *Nature*, **423**, 239 (2003). In an article posted by Lisa Pinsker in *Geotimes* (www. geotimes.org/july03/NN_core.html), Stevenson is quoted as saying, "when I wrote the paper, it was in part a deliberate attempt to shake people up, and it will probably succeed at that independent of whether people finally decide that it's actually an idea that can be carried out." A decade after the appearance of his initial paper, in a *Forbes Magazine* interview (July 31st, 2013), Stevenson indicated that he currently favored some form of self-burying capsule approach (see [36]) for studying the Earth's core.

[38]   For a recent detailed review see Cosimo Bambi, "Astrophysical black holes: A compact pedagogical review," http://arxiv.org/abs/1711.10256. The history and possible properties of black holes have been ably reviewed by Kip Thorne (co-winner of the 2017 Nobel Prize for Physics following the discovery of gravitational waves at LIGO) in the book *The Science of Interstellar* (W. W. Norton & Co., 2014). This text accompanies the November 2014 Warner Brothers movie *Interstellar* by filmmaker Christopher Nolan.

[39]   See I. B. Khriplovich *et al.*, "Passage of small black hole through the Earth: Is it detectable?" http://arxiv.org/abs/0801.4623.

[40]   See Y. Luo *et al.*, "Detectable seismic consequences of the interaction of a primordial black hole with Earth," http://arxiv.org/abs/1203.3806.

[41] See B. E. Zhilyaev, "Singular sources of energy in stars and planets," http://
       arxiv.org/abs/0706.2504. For further possibilities, see also A. P. Trofimenko's
       (imaginative) article: "Black holes in cosmic bodies," *Astrophysics and
       Space Science*, **168**, 277–292 (1990).

[42] The possibility of mini black holes being created at the LHC is entirely predi-
       cated upon the theoretical possibility of there being more than 3 dimen-
       sions to space, but even so, the maximum mass of any new particle and/or
       black hole that might be created in the LHC can be no more than the maxi-
       mum collisional energy of the proton beams: $E_{beam} = m_{BH}c^2$, where c is the
       speed of light. When operating at its maximum capacity $E_{beam} \sim 14$ TeV and
       accordingly, $m_{BH} \sim 10^{-23}$ kg. Even if copious numbers of mini black holes are
       created by the LHC, they will be far too small to accrete any matter, and
       additionally their lifetime against evaporation due to Hawking radiation is
       almost instantaneous. For a detailed discussion see A. F. Ali, M. Faizal &
       M. Khalil, "Absence of black holes at LHC due to gravity's rainbow," *Physics
       Letters B*, **743**, 295–300 (2015). Not only is it impossible for the LHC to pro-
       duce a black hole capable of destroying the Earth, measurements of cosmic
       ray energies, which can be many, many orders of magnitude larger than
       that within reach of the LHC, indicate that black hole production must be
       rare and entirely benign if it happens at all.

[43] The black hole accretion scenario is most often discussed in astrophysical
       terms. Various authors have considered the possible effects of a black hole
       being located at the center of star, and various potentially observable out-
       comes seem possible. In the case of a black hole at the center of a star the
       maximum accretion rate is set according to the accretion flow attaining the
       so-called Eddington luminosity, L_{Edd}, close to the black hole's Schwarzschild
       radius. First introduced by pioneering astrophysicist Arthur Eddington, L_{Edd}
       corresponds to the condition where the pressure support is provided by
       radiation only, giving $L_{Edd} \sim K m_{BH}$, where m_{BH} is the mass of the centrally
       accreting black hole and K is a constant. Under these conditions, the accre-
       tion rate is $dm/dt = L_{Edd}/\eta c^2$, where c is the speed of light and $\eta \sim 0.1$ is an
       efficiency factor. For a black hole of mass of 10^{15} kg, so the maximum
       accretion rate will be of order 1 kilogram of matter per second. For such a
       black hole introduced into the Sun, the settling time would be of order 600
       million years (corresponding to $\beta \sim 10^{-12}$). Intriguingly, it additionally turns
       out, if the Sun could be engineered to generate its luminosity through
       accretion onto a central black hole then its lifetime could, in principle, be
       extended well beyond that which it will otherwise have under standard
       central hydrogen burning conditions (the so-called main sequence

lifetime) — see, for example, the author's book, *Rejuvenating the Sun and Avoiding other Global Catastrophes* (Springer, New York, 2008).

[44] The importance of the $n = 5$ polytrope in explaining red giant structure is described in P. Eggleton, J. Falkner & R. Cannon, "A small contribution to the giant problem," *Monthly Notices of the Royal Astronomical Society*, **298**, 831–834 (1998). The author has also discussed the role of the polytopic approximation in describing advanced phases of stellar evolution in "The formation of red giants," *Astronomy and Astrophysics*, **156**, 391–392 (1986).

[45] In terms of pure fantasy, Edgar Allan Poe made great literary use of Symmes' hollow-Earth idea. Indeed, Poe wrote a prize-winning short story, *MS. Found in a Bottle* (published in the October 19th, 1833, issue of the *Baltimore Saturday Visitor*), and a well-received book, *The Narrative of Arthur Gordon Pym of Nantucket* (1838), both of which involve adventures in the "open" polar seas.

[46] This quotation is taken from Hardy's article "Artists in Space" which appeared in the September–October 1998 issue of *Mercury* magazine. Within the context of illustration and as an aid to the imagination, Hardy is entirely right in his claims. Indeed, while the artistic depiction of some entirely make-believe exoplanetary surface is a total fabrication, it can play a vital role in expanding and exciting the human imagination.

[47] These details are taken from the online version of Yeates' (2004) graduate diploma (University of New South Wales, Australia), *Thought experiments: A cognitive approach*. Additional discussion can be found in James Brown's entry to the Stanford Encyclopedia of Philosophy (http://plato.stanford. edu/entries/thought-experiment). In terms of the philosophical challenge, the problem of thought experiments revolves around the question: "how can we learn about reality by thought processes alone?" Arthur Eddington, see [26], and Roger Penrose, see [3], have adopted the stance that the human mind can, in fact, make contact with the Platonic realm in which ideal forms exist, but what exactly this mind/thought contact process is remains a mystery at the present time. Interestingly, Penrose argues that this "mind bridge" between physical reality and the Platonic realm of ideals is not a process that will be accessible to computational AI — the "bridge" being an inherently non-computational, that is, non-algorithmic, process.

[48] There was nothing especially new in Oldham's experimental procedure, other than he used refined photographic techniques and made accurate theodolite measurements. What is more important with respect to

Oldham's repeat experiment is that his results were accepted for presentation at the 71st meeting of the British Association for the Advancement of Science. In effect this placed the weight and authority of the scientific establishment behind the conclusion that the Earth (and especially the 6-mile stretch of the Bedford Level canal) is indeed, curved. With the British Association's stamp of approval, the scientific community had spoken, and it was to be Oldham's results that were referenced (for many decades) as the definitive proof that the Earth is assuredly a spheroid.

[49]   The names Pluto and Proserpine that Verne attributes to the two stars is pure literary invention, but they appropriately correspond to the King and Queen of the ancient Greek underworld. The dwarf planet Pluto, of course, was not discovered until 1930, while the main-belt asteroid 26 Proserpina was discovered in 1853. Halley did speculate upon the appearance of the aurora (northern lights) in several research article, linking them to some form of water vapor and/or magnetic effluvia, and he even suggested that the inner regions of the Earth might contain the same luminous medium that was visible in the heavens — this matter (in modern terms) corresponding to the luminous clouds of gas and dust observed in the interstellar medium — but he did not argue (as did Leslie) that such light phenomena were due to high compression.

[50]   The term "just a theory" is often invoked to this very day — especially by creationists and those who would deny climate change and global warming. The problem, of course, resides with the public misunderstanding of the word "theory." A scientific theory is an explanation of some specific phenomenon or object that has been verified by repeated experimentation and testing, and which has been accepted as a valid explanation via a peer review process. The lay interpretation of the word "theory" is that it is something fanciful and only a fanciful idea that exists in someone's head. One could argue that Symmes is more honest than Gardiner, since Symmes at least stated his underlying convictions, poorly conceived as they were, in his declaration of 1818 (see Figure 21.3).

[51]   For an oblate spheroid of equatorial radius a and polar radius b, the gravitational acceleration at the poles will be: $g_{pole} \approx (GM/b^2)\,(1 - 6f/5)$, where M is the mass of the sphere and where $0 < f = 1 - (b/a) \ll 1$. At the equator, in contrast, $g_{equator} = (GM/b^2)\,(1 - 7f/5)$. These relationships provide the result that the gravitational acceleration is higher at the poles of an oblate spheroid than at the equator, with $g_{pole}/g_{equ} \approx 1 - f/5$. These results are derived, for example, in A. S. Ramsay's classic textbook, *Newtonian Attraction* (CUP, 1961). Taking the profile of the spheroid to be $R = b(1 + fs^2)$, where $s = \sin\varphi$,

with $0 \leq \varphi \leq \pi$, the gravitational force acting upon a particle moving along the polar axis will be $F(y) = (4\pi/3)G\rho(1 - f/5)y$, with $-b \leq y \leq b$. The gravitational force acting upon a particle moving along an equatorial axis will be $F(x) = (4\pi/3)G\rho(1 + 3f/5)x$, with $-a \leq x \leq a$. In these two special polar and equatorial cases the force acting upon the tunnel-crossing particle is the same as that for a sphere with a constant modification factor Δ such that, $F(r) = (4\pi/3)G\rho\Delta r$, where $\Delta = (1 - f/5)$ in the polar case, and where $\Delta = (1 + 3f/5)$ in the equatorial case. In this manner the period of oscillation will be $T = (3\pi/G\rho\Delta)^{\frac{1}{2}} = T_{\text{sphere}} (1 + f/10)$ in the case of polar motion.

[52] The details of the experiment by Worden and Everitt are given in The Proceedings of the International School of Physics "Enrico Fermi," course 56, held at Verenna July 17–29, 1974 (organized by the Italian Physics Society). Further details are given in the book by Y. T. Cook and A. Chen, *Gravitational Experiments in the Laboratory* (CUP, 2005).

[53] This short story first appeared in *The Magazine of Fantasy & Science Fiction* for April 1966, but has since been republished in various anthologies. Somewhat oddly Dick indicates in his short story that the travel time along *The Fall* is just 17 minutes. This shorter than the "normal" gravity-train travel time of 42 minutes could be achieved, of course, if one employed say rocket or magnetic rail acceleration and deceleration at the beginning and end of the journey.

[54] See the author's book, *Terraforming: The Creating of Habitable Worlds* (Springer, 2009). A detailed discussion is also provided by Martyn Fogg in his extensive technical text, *Terraforming: Engineering Planetary Environments* (SAE Press, 1995). The Kim Stanley Robinson *Mars trilogy* concerning the colonization and terraforming of the Martian surface is also recommended since, although largely a story about the politics and dynamics of a transplanted human society, the books are well researched with respect to potential terraforming techniques.

[55] Asteroids are rich in such important industrial-use minerals as gold, iron, iridium, manganese, nickel, platinum, silver and tungsten. The material mined from asteroids could either be returned to Earth for further processing or engineered to build structures in space. Duncan Forgan (University of Edinburgh) and Martin Elvis (Harvard Smithsonian Center for Astrophysics) have suggested that looking for signs of asteroid mining might be one way of detecting extraterrestrial civilizations. They note that such activities might in principle be observable in the form of "deficits in mineral species, changes in the size distribution of debris and [...] thermal signatures." See http://arxiv.org/abs/1103.5369.

[56] Details concerning the activities at the Vostok Station have been summarized by Nikolay Vasiliev and Pavel Talalay in "Twenty years of drilling the deepest hole in ice," *Scientific Drilling*, No. 11, March 2011. Details concerning the analysis of the first water samples extracted from Lake Vostok are given in a research paper by Yury Shtarkman and co-workers, "Subglacial Lake Vostok (Antarctica) accretion ice contains a diverse set of sequences from aquatic, marine and sediment-inhabiting bacteria and eukarya, published online by *PLOS One* (https://doi.org/10.1371/journal.pone.0067221).

Index

Printed in the United States
By Bookmasters